✧ **Near-Earth Objects** ✧

✧ Near-Earth Objects ✧

Finding Them Before They Find Us

DONALD K. YEOMANS

with a new preface by the author

Princeton University Press
Princeton and Oxford

Second printing, first paperback printing, with a new preface by
the author, 2016

Paperback ISBN: 978-0-691-17333-7

The Library of Congress has cataloged the cloth edition as
follows:
Yeomans, Donald K.
Near-Earth objects : finding them before they
find us / Donald K. Yeomans.
p. cm.
Includes bibliographical references and index.
ISBN 978-0-691-14929-5 (hardcover : alk. paper)
1. Near-Earth objects. 2. Asteroids—Collisions with Earth.
3. Comets—Collisions with Earth. I. Title.
QB651.Y46 2013
523.44—dc23
2012005990

British Library Cataloging-in-Publication Data are available
This book has been composed in Goudy
Printed on acid-free paper. ∞
Printed in the United States of America
3 5 7 9 10 8 6 4 2

✧ CONTENTS ✧

CHAPTER 9

CHAPTER 10

✧ ILLUSTRATIONS ✧

Comets and asteroids represent the building blocks of our early solar system and they likely brought to the early Earth much of the water and carbon-based materials that allowed life to form. Once life did form, continuing collisions of these objects with the Earth modified the evolutionary process allowing only the most adaptable species to evolve further. In a sense, we humans owe our very origins and our current position atop the food chain to these near-Earth objects. Comets and asteroids may one day provide the raw materials for structures in space. They may also provide the water necessary to sustain life and, by breaking down the water into hydrogen and oxygen, the rocket fuel that will power spacecraft traveling within the solar system. These near-Earth objects are far more important than their relatively small sizes would suggest.

Since this book's publication in early 2013, there have been some important events that should be included in a current edition of this book. Two of these events occurred on the same day, February 15, 2013, when one small, known asteroid passed very close to the Earth's surface and another small asteroid collided, unannounced, with the Earth's atmosphere near Chelyabinsk, Russia.

Discovered by the LaSagra observatory in southern Spain, the small asteroid 2012 DA14, now designated (367943) Duende, passed within 4.4 Earth radii (or 27,700 km) of the Earth's surface on February 15, 2013. Asteroid 2012 DA14 passed interior to the Earth's geosynchronous orbital ring, which is located at an altitude of about 35,800 km. Fortunately, none of the communication satellites located in the geosynchronous ring about the Earth, including some that were announcing the arrival of this asteroid, were affected. Observations made near the time of closest approach indicated that this asteroid rotates with a period of nine hours and has a relatively high reflectivity, reflecting about 44 percent of the incident sunlight. Radar observations made near the time of the Earth close approach indicate the object is elongated with dimensions of about 20 x 40 meters. An asteroid of this size would be expected to strike the Earth and cause regional damage with an average interval of about seven hundred years and to get this close to Earth with an average interval of about twenty-five years.

Earlier on the same day (February 15, 2013) an unrelated smaller asteroid collided with the Earth's atmosphere near Chelyabinsk, Russia. This small asteroid caused a large atmospheric fireball when it entered the Earth's atmosphere at high speed and at a shallow angle with respect to the horizon. The atmospheric impact released a tremendous amount of energy at high altitude and produced a shower of fragments of various sizes that fell to the ground as meteorites. Although this small asteroid approached Earth from the direction of the sun, and was hence unobservable prior to impact, the atmospheric fireball it caused was observed not only by Russian automobile video cameras and an international network of low frequency infrasound detectors, but also by U.S. Government sensors in Earth orbit. As a result, the details of the impact have become clear. The altitude of the fireball event was approximately twenty-three kilometers (fourteen miles) and its total impact energy was equivalent to about 440 thousand tons of TNT explosives. Because of the nearly ninety second delay between the fireball's optical flash and the arrival of the fireball shock wave, many citizens in the city of Chelyabinsk rushed to their windows when the optical flash occurred and then were hit with flying glass when the shock wave arrived. Approximately 1500 citizens required medical treatment, but fortunately there were no fatalities.

The meteorites recovered from the Chelyabinsk fireball are reported to be so-called ordinary chondrites, which have a typical density of about 3.6 grams per cubic centimeter. Water has a density of one gram per cubic centimeter. Given the total energy of about 440 thousand tons of TNT explosives, the approximate diameter of the Chelyabinsk asteroid would have been about eighteen meters in diameter, and its pre-impact mass would be roughly eleven thousand tons.

The path of the Chelyabinsk asteroid, prior to impact, was that of a typical near-Earth asteroid traveling about the sun. It reached its greatest distance from the sun within the asteroid belt between the orbits of Mars and Jupiter and its smallest distance from the sun near the orbit of Venus, inside the orbit of Earth. On February 15, 2013, the Earth and the asteroid arrived at the same point at the same time, creating the impressive Chelyabinsk impact event.

Though Asteroid 2012 DA14 made its very close flyby of the Earth only about sixteen hours after the Russian fireball event, there is no connection whatsoever between these two events. First of all, the two objects approached the Earth from completely different directions, and had entirely different orbits about the Sun.

A second reason we know the two asteroids approaching Earth on February 15 were unrelated is their different compositions. Telescopic spectral data do not support any physical connection between Asteroid 2012 DA14 and the Chelyabinsk meteorites. Asteroid 2012 DA14 displays spectral colors in the "L-class" of asteroid classification, which suggests a carbon-dominated composition with abundant calcium and aluminum-rich inclusions. On the other hand, meteorite fragments being recovered from the Chelyabinsk fireball event are reported as silicate-rich ordinary chondrites, a completely different and unrelated class of meteorites.

There are thought to be more than a million objects the size of the Chelyabinsk asteroid in near-Earth space, but we have found only a small fraction of this population. Because continuing collisions between relatively large asteroids produce multiple collision fragments, or smaller asteroids, there are far more small near-Earth asteroids than large ones. There are thought to be nearly one thousand near-Earth asteroids one kilometer (0.6 miles) or larger in diameter, more than a million the size of the Chelyabinsk impacting asteroid, and one hundred million near-Earth asteroids about the size of an automobile (i.e., four meters in diameter). Hence, whereas the Earth can expect to be struck by a one-kilometer-sized asteroid only every five hundred thousand years on average (albeit with catastrophic results), the average impact interval for a Chelyabinsk-sized object is about once every fifty years and a few times each year for asteroids the size of an automobile. These latter objects deform or "pancake" under the pressure of the Earth's atmosphere, and unless they have a rare nickel-iron composition, they would not be expected to survive passage through the Earth's atmosphere. The first asteroid discovered in 2014, 2014 AA, was three to four meters in size and it impacted the Earth's atmosphere over the South Pacific Ocean on New Year's Day. It would have gone completely unnoticed were it not for highly sensitive Earth-based and space-based sensors that are designed to monitor atomic bomb tests.

Each year, the near-Earth asteroid discovery rate increases as the detection technology and techniques improve. In 1975 only about two near-Earth asteroids were discovered each year. In 1990, this discovery rate increased to about two per month, and in 2016, the rate is nearly two hundred per month. Until relatively recently, we were completely unaware of this swarm of potentially hazardous asteroids in the Earth's neighborhood. Today, we'll need to redouble our efforts to find these large near-Earth asteroids well in advance of a threatening Earth encounter. Our very survival depends upon our knowing of these potential threats far enough in advance that counter measures, or asteroid deflections, can be undertaken.

We'll need to find them—before they find us.

✧ PREFACE ✧

Before the relatively recent discovery of a great population of asteroids in the Earth's neighborhood, a work on the asteroids and comets that make up the so-called near-Earth objects would have been more of a pamphlet rather than a book of this size.

The show-off comets, with their enormous gas and dust tails, have been recorded for millennia. They were feared as mysterious apparitions presaging disasters by the ancient Greeks and Chinese and as fireballs thrown at a sinful Earth from the right hand of an avenging God during the church-oriented Middle Ages. In late 1694, Edmond Halley, who first correctly predicted the return of the comet that bears his name, speculated, "comet impacts may have formed the vast depression of the Caspian sea and other great lakes in the world." In 1822, the British poet Lord Byron imagined a time when men would have to defend Earth from these celestial miscreants.

> Who knows whether, when a comet shall approach this globe to destroy it, as it often has been and will be destroyed, men will not tear rocks from their foundations by means of steam, and hurl mountains, as the giants are said to have done, against the flaming mass?—and then we shall have traditions of Titans again, and of wars with Heaven.[1]

Although the show-off comets in the inner solar system are impressive, it is the far more numerous asteroids in the Earth's neighborhood that should have been feared as they represent the most frequent threats to Earth. However, the threats from these near-Earth asteroids have

[1] Lord Byron in E. J. Lovell, Jr., ed., *Medwin's "Conversations of Lord Byron"* (Princeton: Princeton University Press, 1966), 188.

only recently been realized, having repeatedly slipped by Earth for eons without notice. Eros was the first near-Earth asteroid to be discovered in 1898, and it was another thirteen years until the discovery of the second one named Albert. Albert was only observed for a month and then lost for nearly a century before being rediscovered in 2000. By 1950, only thirteen near-Earth asteroids had been discovered—all by accident when astronomers were observing other objects in the night sky. Deliberate photographic searches for near-Earth asteroids begun only in the 1970s and 1980s found a few more so that by 1990, there were 134 known near-Earth asteroids. Rapid progress in discovering near-Earth asteroids only began in the 1990s when deliberate, systematic search programs, using electronic CCD detectors and digital computer processing rather than photographic techniques, got under way. Largely as a result of higher sensitivity telescopic survey programs supported by NASA, there were more than 8,800 known near-Earth asteroids in early 2012 and they continue to be discovered at a surprising rate. Whereas it took sixty-two years between the first discovery of a near-Earth asteroid in 1898 and the twentieth discovery in 1960, today's search programs discover about twenty near-Earth asteroids in a single week.

While Earth impacts by large near-Earth objects are very low probability events, they are of very high consequence. While no one can claim knowledge of a single person being killed by a near-Earth object, the evidence for major impact events taking place in Earth's history is unmistakable. However, one could claim that even over very long time intervals, the average annual number of those killed by Earth impacts is comparable with the fatalities due to shark attacks or fireworks accidents, and there are far more fatalities due to, say, automobile accidents. So why be concerned? The point is that near-Earth object impacts, unlike shark attacks, fireworks accidents, automobile accidents, and any number of other more familiar disasters, have the capacity to wipe out an entire civilization in a single blow.

✧ **ACKNOWLEDGMENTS** ✧

I doubt that any book is the work of a single author. This one certainly isn't. I received, and very much appreciate, the comments provided by several authorities in planetary science, many of whom could have written this entire volume on their own. My colleagues at the Jet Propulsion Laboratory (JPL), including Alan Chamberlin, Steve Chesley, Paul Chodas, and Jon Giorgini reviewed portions of the book. These gentlemen provide the technical magic for monitoring the motions of all asteroids and comets in the Earth's neighborhood. A good portion of this book tells the story they have written. Sections of the book were read and commented upon by noted scientist Mark Boslough (Sandia Labs), David Dearborn (Lawrence Livermore National Laboratory), and ex-astronauts Tom Jones and Rusty Schweickart. Their help is very much appreciated. Lindley Johnson, the program executive for the near-Earth object program at NASA headquarters, provided constructive comments on the entire book. NASA is very fortunate to have someone of his caliber running the near-Earth object program. David Morrison (Ames Research Center and the SETI Institute) and Clark Chapman (Southwest Research Institute), who were both instrumental in focusing early attention on the issues involving near-Earth objects, provided several suggestions and improvements for the entire book as did Dan Scheeres (University of Colorado at Boulder). Dan is everyone's go-to guy for questions involving the dynamics of comets and asteroids. You'll hear more about these people in the coming pages.

Thanks are also due to Ingrid Gnerlich, Senior Editor at Princeton University Press, for initially suggesting this book, and to production editor Debbie Tegarden for seeing it through to publication. They were quick to provide patient and reasonable responses to my many, many questions.

I'd like to acknowledge the support of my daughter, Sarah, an archaeologist, and my son, Keith, an attorney. Both are well loved and I'm very proud of them. I should also mention "wee Henry," our first grandchild, who was born when these words were written in October 2011. Who knows what wonders and technological magic he'll witness in his lifetime. My wife, Laurie, always understood and accepted that my JPL position, as well as work on this book, would translate into a fair number of working evenings and weekends. She's been my love for more than four decades.

Near-Earth Objects

✧ CHAPTER 1 ✧

Earth's Closest Neighbors

The dinosaurs became extinct because they
didn't have a space program.
—Larry Niven

Michelle Knapp and Her 1980 Chevrolet Malibu

Let me introduce Michelle Knapp of Peekskill, New York, and her
1980 Chevy Malibu sedan. On a rainy Friday night, October 9, 1992,
just before 8:00 PM, Michelle, an eighteen-year-old high school senior,
heard a loud crash in her driveway and raced outside to discover the
rear end of her automobile had been completely destroyed by a football-
sized rock. The twenty-seven-pound projectile had punched completely through the trunk, just missing the gas tank.

As unlikely as it sounds, a fragment of a near-Earth asteroid that had
collided with Earth destroyed Michelle's car. The fiery trail of the initial, Volkswagen-sized, near-Earth asteroid was first seen over West
Virginia appearing with a greenish hue and brighter than the full
Moon. Due to the forces of the atmospheric resistance, the asteroid
fragmented into more than seventy pieces while traveling northeast for
more than forty seconds over Pennsylvania and then New York. The
only known surviving fragment came to a full stop underneath
Michelle's Chevy Malibu. Countless people, many of whom were
watching high school football games that Friday evening, observed the
fiery train of fragments in the Pennsylvania and New York skies.

F<small>IGURE</small> 1.1. Michelle Knapp and her 1980 Chevy Malibu. After a small asteroid entered the Earth's atmosphere on October 9, 1992, it caused a spectacular fireball event traveling northeast over parts of West Virginia, Pennsylvania, and New York before coming to rest underneath Michelle's 1980 Chevrolet Malibu sedan.
Source: Courtesy of John E. Bortle, W. R. Brooks Observatory.

Although Michelle's automobile insurance company refused to pay for her car's damages, claiming that it was an act of God, she got the last laugh by selling the so-called Peekskill meteorite and the twelve-year-old Chevy to a consortium of three meteorite collectors for $69,000.

On a daily basis, at least one hundred tons of interplanetary material rain down upon the Earth's atmosphere, but most of it is in the form of very small dust particles or very small stones. Much of this dust and sand grain–sized material, the debris from active comets, can be seen on almost any clear, dark night as meteors or shooting stars. Larger basketball-sized rocks rain down upon the Earth daily and while they can cause impressive fireball events, our atmosphere prevents almost all of them from reaching the ground intact. Volkswagen-sized asteroids, like the one that fragmented and caused the Peekskill meteorite, strike

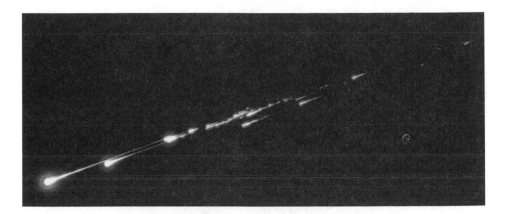

FIGURE 1.2. Because of the atmospheric pressure, the small asteroid that created the Peekskill fireball event fragmented into more than seventy pieces in the Earth's atmosphere. This image was taken from Mansion Park football stadium in Altoona, Pennsylvania. Only a single meteorite was located on the ground.
Source: S. Eichmiller of the *Altoona Mirror*.

the Earth's atmosphere every six months or so on average. At this point, you may be incredulous because you may not have ever seen a fireball and probably not a major fireball like the Peekskill event. But the vast majority of the Earth's surface is either ocean or unpopulated and besides, how often do you monitor the skies all night? Department of Defense satellites do, however, continuously monitor the skies and detect fireballs, as they look downward, twenty-four hours each day, to provide alerts of possible missile launch events and nuclear explosions.

Mr. S. B. Semenov Gets Blown off His Porch

Allow me to introduce Mr. S. B. Semenov, who was an eyewitness to a larger Earth-impacting near-Earth asteroid on June 30, 1908, in a remote region of Russian Siberia called Tunguska.

Mr. Semenov, a farmer, was sitting at a trading post when he noticed what appeared to be a fire high and wide over the local forest. A loud and strong shock ensued, blowing him a few meters off the trading post

FIGURE 1.3. Mr. S. B. Semenov, eyewitness to the Tunguska blast of June 30, 1908. Although he was sixty-five kilometers distant from the blast site, he was thrown off a porch and related that his shirt felt like it was on fire.
Source: Courtesy of E. L. Krinov.

porch. He noted that the heat from the blast felt like his shirt was on fire, even though he was located about sixty-five kilometers south of ground zero. Although a wide variety of suggestions have been made to explain the Tunguska event, including the absurd notions that the blast was due to a UFO crash or an overzealous signal greeting by aliens, by far the most likely cause of the Tunguska blast was an atmospheric impact by a near-Earth asteroid. Most likely, a thirty-meter-sized asteroid entered the Earth's atmosphere and reached an altitude of about eight kilometers before the atmospheric pressure in front of the stony object "pancaked" the rock, causing it to explode above the forest floor. A tremendous blast wave then continued to the surface, and an area that spanned some two thousand square kilometers of forest, involving millions of trees, was leveled. But since the stony asteroid itself disintegrated in the air blast explosion and did not reach the ground, no crater was evident on the forest floor and no sizable meteorites were left near ground zero. Current estimates place the energy of the event at about four million tons (four megatons) of TNT high explosives.[1] Considering there are more than a million of these asteroids, thirty meters and larger, in the Earth's neighborhood, one would expect an asteroid of this size or larger to strike the Earth every few hundred years. Thirty meters, or Tunguska-sized, is about the minimum diameter of an Earth impactor that could cause significant ground damage. In general, smaller stony objects would not be expected to survive passage through the Earth's atmosphere.

The Dinosaurs Check Out Early

At the upper end of the near-Earth object sizes are about one thousand asteroids larger than one kilometer in diameter, and an Earth impact by one of these asteroids would be capable of causing global devastation. Fortunately asteroids of this size would not be expected to strike Earth but every seven hundred thousand years on average, and NASA's ongoing search programs have already found more than 90 percent of this population. None of them represents a credible threat during the next century. The very largest near-Earth objects are as large as ten kilometers in diameter. Sixty-five million years ago, one of them killed much of the land and sea flora and fauna along with almost all of the large vertebrates on land and at sea. Most species were exterminated. Crater evidence for this major extinction event has been found near Chicxulub on the edge of the Mexican Yucatán peninsula. A ten-kilometer impactor could cause a so-called extinction event because it

[1] Chapter 8 goes into more detail on the Tunguska event of 1908.

would subject the Earth to global firestorms, severe acid rain, and the darkening of the skies with soot and impact-created debris. The resultant loss of photosynthesis would cause plants to die along with the animals and marine life that depend upon these plants for food. After more than a 160-million-year run, the large land dinosaurs could not survive this impact event because of the complete disruption of their food chain. An Earth impact by a ten-kilometer-sized near-Earth object would create an impact of unimaginable energy—equivalent to some fifty million megatons of TNT. To put that in perspective, that amount of energy would be equivalent to a Hiroshima-type nuclear blast every second for about 120 years! This underscores the statement that while large near-Earth object impacts are very rare, they are extremely high-consequence events capable of ending civilization as we know it.[2]

One of NASA's goals is to discover and track the vast majority of the relatively large near-Earth asteroids and comets that would be capable of threatening Earth or causing a local or regional disaster. If we find them early enough, we now have the technology to deal with them. For example, a massive spacecraft could be directed to purposely run into an Earth-threatening asteroid of modest size to slow it down and alter its trajectory just enough so that it would no longer threaten Earth. As it turns out, the dinosaurs became extinct because they didn't have a space program.

Just What Are Comets, Asteroids, Meteoroids, Meteors, and Meteorites?

In interplanetary space, a large rocky body in orbit about the Sun is referred to as an asteroid or sometimes a minor planet. They are inactive and, unless struck by another nearby asteroid, they do not shed material like their cousins the comets. Comets differ from most asteroids in that they are icy dirtballs. When they approach the Sun, their ices (mostly water ice) are warmed by the Sun so they begin to

---

[2] Chapter 4 goes into more detail on the extinction event that took place sixty-five million years ago.

vaporize and release the dust particles that were once embedded in their ices. Inactive comets that have exhausted their supply of ices near the surface or have their ices covered and insulated by more rocky material are no longer termed comets but asteroids. The only real difference between a comet and an asteroid is that comets, when near the Sun, are actively losing their ices and dust, often causing a highly visible trail of dust and gas, and asteroids are not. Even objects that are icy, like some asteroids and other bodies in the outer solar system, are just classified as asteroids since they do not get close enough to the Sun for their ices to vaporize. They are not active and hence they are not comets. Since the physical makeup of an active icy body (comet) near the Sun can be identical to an inactive icy body (asteroid) that is farther from the Sun, the line between comets and asteroids cannot be clearly drawn.

Small collision fragments of inactive asteroids or the dusty debris from active comets in orbit about the Sun are called meteoroids if their sizes are between ten microns, the width of a cotton fiber, and one meter in diameter. Once a tiny meteoroid enters the Earth's atmosphere and vaporizes due to atmospheric friction, it emits light that causes a meteor or "shooting star." Almost all meteors are due to cometary sand- or pebble-sized particles while small asteroids or large meteoroids can cause much brighter fireball events in the Earth's atmosphere. Fireball events can range in brightness from just brighter than the brightest planets to events that briefly rival the Sun. If a fragment of the impacting body survives its passage through the Earth's atmosphere and lands upon the Earth's surface, it is then called a meteorite.

The Near-Earth and Potentially Hazardous Objects

Astronomers refer to the approximate average distance between the Sun and Earth as an astronomical unit (AU), which is a distance of about 150 million kilometers or 93 million miles. Near-Earth objects are simply defined as comets and asteroids that approach the Sun to within 1.3 AU so therefore they can also approach the Earth's orbit to within 0.3 AU if their orbits are close to the same plane as that of the

Earth. The so-called potentially hazardous objects are a subset of the near-Earth objects that approach the Earth's orbit to within 0.05 AU, which is roughly the distance that a near-Earth object's trajectory can be gravitationally altered by a single planetary encounter. These objects would have to be about 30 meters in size or larger to cause significant damage at the Earth's surface.

Although sizable near-Earth asteroids outnumber near-Earth comets by more than one hundred to one, the solid nuclei of comets may provide the very largest impactors like the one that took out the dinosaurs. Cometary debris is also the source of most tiny meteoroid particles and meteors. Many comets generate meteoroid streams when their icy cometary nuclei pass near the Sun, begin to vaporize, and release the dust, sand-sized particles, and fragile clumps that were once embedded in the cometary ices. These meteoroid particles then follow in the wake of the parent comet. When the Earth, in its orbit about the Sun, runs into this dusty debris from active comets, meteor showers can be observed. Sometimes hundreds and even thousands of meteors, or shooting stars, can be seen within an hour when the Earth collides with a particularly dense band of meteoroids. The annual August Perseid showers occur when the Earth runs into small particles from comet Swift-Tuttle while the November Leonid showers are caused by the debris from comet Tempel-Tuttle.

Occasional collisions between asteroids in the main asteroid belt located between the orbits of Mars and Jupiter create asteroid fragments, and it is these fragments that are the sources of most near-Earth objects. These fragments are also the source of most Earth-impact events and the meteorites that have survived these violent collisions with Earth. As time goes by, asteroids colliding with one another in the inner planetary region produce more and more smaller fragments while reducing the number of larger ones. As a result, we are fortunate that the vast majority of near-Earth objects that collide with Earth are far too small to survive the Earth's atmosphere and there are relatively few objects in near-Earth space that are large enough to cause global consequences upon impact with the Earth.

Because they are readily available for study, many meteorites have already been subjected to detailed chemical and physical analyses in laboratories. If particular asteroids can be identified as the sources for some of the well-studied meteorites, a detailed knowledge of the meteorite's composition and structure will provide important information on the chemical mixture and conditions from which the parent asteroid formed 4.6 billion years ago.

The Orbits of Near-Earth Objects

There are four orbital classes of near-Earth asteroids with membership in each class being determined by a particular asteroid's orbital characteristics compared to the Earth's orbit. The Earth's orbit about the Sun is nearly circular—but not quite. An orbit's eccentricity (e) is a measure of the orbit's departure from a circle. For a circular orbit, e = 0, and as the orbit gets more and more elongated, or eccentric, the eccentricity increases toward one. An open parabolic orbit has an eccentricity of one and a hyperbolic orbit greater than one. Earth's orbital eccentricity is 0.0167. Earth reaches its closest point to the Sun (perihelion) in early January at a heliocentric distance of about 0.983 AU while it

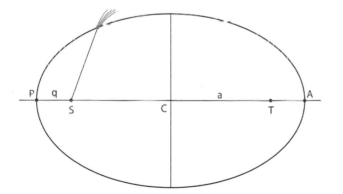

FIGURE 1.4. Some orbital characteristics. The Sun (S) is located at one of the two foci of the asteroid's elliptical orbit. Half the long, or major, axis of the ellipse is termed the semi-major axis (CP). The ratio of the distances CS to CP is the orbital eccentricity. The perihelion distance (SP) is usually denoted as "q" while the aphelion distance (SA) is denoted "Q."

TABLE 1.1. *Definitions of near-Earth objects*

Asteroid	A relatively small, inactive, (usually) rocky body orbiting the Sun
Comet	A relatively small, at times active object whose ices can vaporize in sunlight, forming an atmosphere (coma) of dust and gas and, sometimes, tails of dust and gas
Meteoroid	A small particle from a comet or asteroid that is orbiting the Sun and is less than one meter in extent
Meteor	The atmospheric light phenomenon that results when a meteoroid enters the Earth's atmosphere and vaporizes
Fireball	An event that is brighter than a meteor ranging in brightness from just brighter than the brightest planets to approaching the brightness of the Sun
Meteorite	A meteoroid that survives its passage through the Earth's atmosphere and lands upon the Earth's surface
Near-Earth Object	A comet or asteroid that can approach the Sun to within 1.3 AU. In order for a body to be classified as a near-Earth object, it must have an orbital period less than 200 years
Potentially Hazardous Object	A comet or asteroid that can approach the Earth's orbit within 0.05 AU or about 7.5 million kilometers and is large enough to cause impact damage

reaches its farthest point from the Sun (aphelion) in early July at a distance of about 1.017 AU.[3] An object's perihelion and aphelion distances are usually denoted with the letters "q" and "Q," respectively. The distance of the longest axis of a celestial object's orbit is termed its major axis and, not surprisingly, the semi-major axis (a) is one-half this

---

[3] The Earth is closest to the sun when it is winter in the northern hemisphere and farthest from the sun when it is summer there. This underscores the fact that the seasons have more to do with the orientation of the Earth's rotation pole than its distance from the sun. During summer in the northern hemisphere, the Earth's rotation pole is oriented more toward the sun than in winter so the sun is more nearly overhead with more concentrated sunlight than in the winter months.

Near Earth Object Orbit Classes	Orbit Criteria	
Amors	Earth approaching asteroids with orbits exterior to Earth's but interior to the orbit of Mars. Hence their semi-major axes are larger than 1.0 AU and their perihelia are between 1.017 and 1.3 AU.	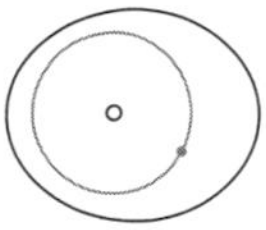
Apollos	Earth orbit crossing asteroids with semi-major axes larger than Earth's (a is larger than 1.0 AU) and with perihelia less than 1.017 AU.	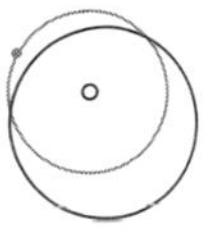
Atens	Earth orbit crossing asteroids with their semi-major axes smaller than Earth's and their aphelia larger than 0.983 AU.	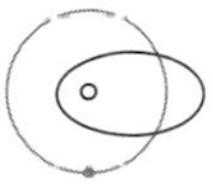
Atiras	Asteroids with their orbits contained entirely within that of the Earth. Hence their semi-major axes are less than 1 AU and their aphelia are less than 0.983 AU.	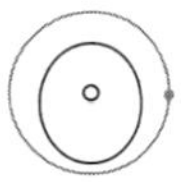

FIGURE 1.5. The four heliocentric orbit classifications for the near-Earth asteroids.

distance. These terms are mathematically related to one another and for closed elliptical orbits, the perihelion distance $q=a(1-e)$ and the aphelion distance $Q=a(1+e)$.

In 1619, the German astronomer Johannes Kepler put forward a fundamental law of planetary motion, which can be expressed as the square of the orbital period (P), in years, is equal to the orbital semi-major axis (a) in AU raised to the power of three. For example, a near-Earth object with a semi-major axis of 2 AU would have an orbital period of 2.8 years (i.e., $2.8 \times 2.8 = 2 \times 2 \times 2$). Some near-Earth asteroids and many comets have orbits that are highly inclined to the plane of the Earth's orbit about the Sun, the plane called the ecliptic. The highest orbital inclination for a planet is seven degrees for Mercury.

The four groups of near-Earth asteroid orbits have real bodies as their namesakes. There is the Earth orbit crossing Apollo group named after asteroid (1862) Apollo, the Earth orbit approaching Amor group named after (1221) Amor, the Earth orbit crossing Aten group, named after (2062) Aten, whose semi-major axes are smaller than that of the Earth's orbit, and finally the Atira group, named after (163693) Atira, whose orbits lie entirely within the Earth's orbit. It is some members of the Aten and Atira orbit groups that are most similar to the Earth's orbit and hence they are, at the same time, the most easily reached asteroids using spacecraft and the ones most likely to run into the Earth.

Rock Stars: Naming Asteroids

There are more than half a million known asteroids in the region between the orbits of Mars and Jupiter and several thousand known near-Earth objects of various sizes in the Earth's neighborhood. These numbers are rapidly growing as more and more of them are discovered. Currently, more than three thousand asteroids a month are being discovered and dozens of these monthly discoveries are in the near-Earth object population.

When a comet is discovered, it is usually named for the discoverer or the discovery program and given a temporary designation to indicate the year and time of year of the discovery. The year of discovery is followed by a letter to denote the half-month during which the discovery took place (I and Z are not used, reducing the alphabet to twenty-four letters). A comet designated 2011 A2 would indicate the comet was the second comet (2) discovered during the first half of January (A) in 2011. Periodic comet Swift-Tuttle was discovered on July 16 by Lewis Swift in Marathon, New York, and independently three days later by Horace Tuttle at Harvard College.[4] Its designation is then P/1862 O1 because it was the first comet discovered in the second half of July 1862. The "P" designates it as a periodic comet that returns on a regular basis. Once there are a sufficient number of observations to allow a secure orbit to be determined, then periodic comets are permanently numbered sequentially (e.g., 1P/Halley, 109P/Swift-Tuttle).

---

[4] Horace Tuttle (1837–1923), a co-discoverer on both of these comets, was during his checkered career a successful astronomer and a Civil War hero; he was also convicted of embezzlement by the U.S. Navy and dismissed for "scandalous conduct tending to the destruction of good morals." See D. K. Yeomans, *Comets: A Chronological History of Observation, Science, Myth and Folklore* (New York: John Wiley and Sons, 1991), 238–39.

In a similar fashion, asteroids are first given preliminary designations according to the year, half-month, and discovery number within that particular half-month. Then, once the asteroid has been well observed and its orbit is secure, the asteroid is given a sequential number. Once an asteroid is given a permanent number, the discoverer has the privilege to name it after a person, place, or thing of his or her choice. Politicians and military figures who have not been dead for one hundred years are ground-ruled out. So are pets and attempts to sell an asteroid name.[5] These "rules" have not been in place that long and sometimes they are not enforced too rigorously, so there are, in fact, asteroids named after three dogs and a cat. With so many asteroid names, there are ample opportunities for unabashed silliness. For example, one could string together the names of the asteroids whose numbers are 9007, 673, 449, 848, and 1136 and arrive at "James Bond Edda Hamburga Inna Mercedes."

There is a certain amount of immortality associated with having your name on an asteroid because it will outlast you by millions of years. Asteroid 2956 Yeomans, a nine-kilometer-sized silicate rock, will be circling the Sun between the orbits of Mars and Jupiter long after this author is dead and gone. Asteroid names are often associated with important scientists, significant artists, adored musicians, and classical composers like (2001) Einstein, (6701) Warhol, (8749) Beatles, and (1814) Bach.

The Importance of Near-Earth Objects

The relatively small comets and asteroids that make up the near-Earth object population are not just poor cousins to the much larger planets. They are the remnants of the planetary formation process itself, and as

---

[5] Asteroid names are approved by the Committee for Small Body Nomenclature of the International Astronomical Union. They are official and permanent. However, for a fee, a number of companies will sell you a fancy certificate that claims to somehow associate your name with a star, but these names are not official and not recognized by any international scientific organization. Apart from a few bright, historic stars that carry official names like Sirius, Betelgeuse, and Rigel, no stars have names, just numerical designations.

the least changed bodies from that process they offer clues to the chemical and thermal conditions under which the planets formed some 4.6 billion years ago. They also likely laid down a veneer of the carbon-based materials and water to the early Earth, thus allowing life to form. Subsequent collisions punctuated evolution, allowing only the most adaptable species, including the mammals, to evolve further. In some sense we owe our very existence and our position atop the food chain to these objects that are, from time to time, our very closest neighbors.

Near-Earth objects also represent potential targets for future human exploration and resources for interplanetary habitats. They are rich in metals and minerals that can be used to build interplanetary shelters and habitats. Their hydrated, or clay, minerals and ices can be used to provide life-giving water and the water can be broken down into hydrogen and oxygen, the most efficient form of rocket fuel. These near-Earth objects may one day act as interplanetary watering holes and fueling stations.

In April 2010 President Obama asked NASA to identify a human exploration mission to a near-Earth asteroid as a stepping-stone mission for the human exploration of Mars. The necessary space technologies and mission risks associated with a several-year human mission to Mars could be tested with a much safer and relatively quick visit to a nearby asteroid. Ironically, the easiest near-Earth asteroids to reach for human exploration are also in orbits that make them the most dangerous in terms of potential future Earth close approaches. The knowledge gained on the asteroid's structure and composition resulting from the human exploration of a near-Earth asteroid could be put to good use, not only for a future human exploration of Mars but also if one of these asteroids were found to be on an Earth-impact threatening trajectory.

CHAPTER 2

The Solar System's Origin

The Classical View

> This is a present from a small, distant world, a token of
> our sounds, our science, our images, our music, our
> thoughts and our feelings. We are attempting to survive
> our time so we may live into yours.
> —President Jimmy Carter

In the Beginning

On September 5, 1977, the Voyager I spacecraft was launched from Cape Canaveral, Florida, and began its audacious journey of planetary exploration. Its suite of scientific instruments included an imaging camera and devices to measure the atmospheres and environments of the giant gas planets Jupiter and Saturn. The spacecraft carried a gold-anodized aluminum record with an engraved schematic diagram showing the Earth's location in our solar system and simply drawn, nude male and female human figures that were very controversial at the time. The record included messages from U.S. president Jimmy Carter and UN secretary-general Kurt Waldheim, spoken greetings in fifty-five languages, and a variety of Earth's natural sounds like wind and thunder.[1] In addition, there were twenty-seven selections of musical recordings from diverse cultures including the first movement of Johann Sebastian Bach's Brandenburg Concerto Number 2 and the 1958 rock and roll classic "Johnny B. Goode" by Chuck Berry. The latter was rated number one in *Rolling Stone* magazine's 2008 list of the one hundred greatest guitar songs of all time.

With an outward velocity of 17 kilometers per second, roughly seventeen times the velocity of a bullet from a high-speed rifle, the Voyager I spacecraft is currently moving away from the Sun faster than any other spacecraft. Let's ride along with Johann Sebastian Bach and Chuck Berry to get an idea of the distances between major solar system bodies beyond our Earth. After its launch in September 1977, Voyager I took less than three months to reach the orbital distance of Mars (1.5 AU from the Sun), although Mars itself was not close to the spacecraft at that time. Just three months later, the spacecraft was passing through the inner region of the asteroid belt (2.5 AU) that is located between the orbits of Mars and Jupiter (5.2 AU). The vast majority of the near-Earth asteroids evolve into the Earth's neighborhood from their birthplace near the inner asteroid belt because of the gravitational nudging of Jupiter and Saturn. In March 1979, Voyager I reached Jupiter with its retinue of Trojan asteroids that both lead and follow Jupiter around the Sun. At Jupiter, the spacecraft passed within 4 Jupiter radii of the planet's center, and after taking advantage of a slingshot-like, gravity assist, our intrepid travelers Johann and Chuck set out upon the twenty-month trip to Saturn, whose mean distance from the Sun is 9.5 AU. A successful flyby to within four thousand kilometers of Saturn's largest moon, Titan, in November 1980 required that the spacecraft leave the ecliptic plane of the planets, and hence no more planetary flybys were possible. However, the spacecraft passed the orbital

---

[1] The contents of the records on each of the Voyager I and Voyager II spacecraft were selected for NASA by a committee chaired by Carl Sagan of Cornell University. In late 2011, the Voyager I and Voyager II spacecraft were at distances of 118 AU and 97 AU from the sun, respectively, and both were still communicating with the giant antennas of NASA's Deep Space Network—although the signals, traveling at the speed of light, take more than thirteen hours to make the trip.

distance of Uranus (19.2 AU) in April 1984 and the orbital distance of Neptune (30.1 AU) three years later in April 1987.

Just beyond the orbit of Neptune, in the region between 35 and 50 AU from the Sun, is the so-called Kuiper belt (actually a flattened torus) of icy small bodies.[2] The dwarf planet Pluto is one of the largest bodies within this group, having been demoted from full planetary status by a group of heartless international astronomers in 2006.[3] The Voyager spacecraft completed its passage by this doughnut-shaped belt of inactive, comet-like bodies in late 1992. Short-period comets can evolve into near-Earth objects from the Kuiper belt or perhaps more likely from the so-called scattered disk objects whose orbits can extend out well beyond the traditional 50 AU boundary for Kuiper belt objects to 300 AU or more. Near their perihelia, Neptune can perturb scattered disk objects and have their perihelia reduced. The so-called centaurs are a population of icy bodies between the orbits of Jupiter and Neptune that are thought to be the intermediate stage in a comet's migration from the scattered disk to the inner solar system. Uranus,

[2] The Kuiper belt was named for the Dutch American astronomer Gerard Kuiper (1905–73), who, in 1951, suggested the notion of a collection of bodies beyond Pluto. However, he made this suggestion based upon the ability of a massive Pluto to gravitationally scatter objects outward toward the Oort cloud of comets. Since Pluto's mass is now known to be far less than Kuiper thought, the so-called Kuiper belt that we now refer to is not really the belt that Kuiper suggested. A suggestion of a collection of bodies beyond Neptune had been put forward eight years earlier by the British astronomer Kenneth Edgeworth, and some astronomers prefer to use the term "Edgeworth-Kuiper belt." The Uruguayan planetary scientist Julio Fernández clearly pointed out in 1980 that comets with short orbital periods likely originated from a flattened disk of icy bodies beyond Neptune so if life were fair, we would now refer to the Kuiper belt as the Fernández belt. As pointed out by David Jewitt, the name "Kuiper belt" then follows Stigler's law, which states, "no scientific discovery is named after its original discoverer." In an apparent confirmation of this tongue-in-cheek law, the University of Chicago statistics professor Stephen Stigler attributed Stigler's law to the sociologist Robert K. Merton. It was Dave Jewitt and Jane Luu who discovered the first Kuiper belt object (other than Pluto) in 1992 from an observatory on Mauna Kea, Hawaii. More than one thousand of them have been discovered, with some comparable in size to Pluto. The term "transneptunian objects" is often used to describe the icy bodies, in the Kuiper belt and the scattered disk, that orbit the sun beyond the orbital distance of Neptune.

[3] During the 2006 General Assembly of the International Astronomical Union in Prague, a vote of the part of the membership still attending was taken to change the status of poor Pluto from a planet to a dwarf planet. Until this vote, Pluto had been the ninth planet in our solar system since its discovery in 1930 by Clyde Tombaugh at the Lowell Observatory in Flagstaff, Arizona. This reclassification was controversial; a few astronomers refuse to refer to Pluto as a dwarf planet.

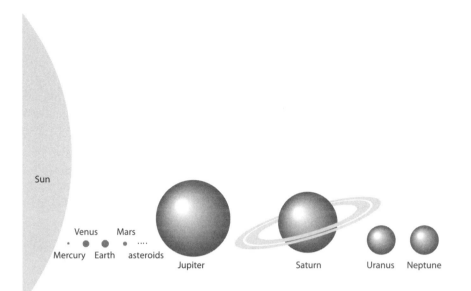

FIGURE 2.1. The eight major planets range in size from the smallest, most interior planet, Mercury, with a diameter of 4,879 kilometers to the largest, Jupiter, whose equatorial diameter is more than twenty-nine times Mercury's size with a diameter of about 143,000 kilometers. For comparison, the Earth's equatorial diameter is 12,756 kilometers. The approximate relative sizes of the planets are evident but the distances between the planets are not to scale.

then Saturn, and finally Jupiter in turn perturb centaurs before they become short-period comets under the control of Jupiter.

Some parochial solar scientists have referred to the region just beyond the Kuiper belt as the edge of our solar system, but at their current speed Johann and Chuck would have to continue their journey for another twenty-eight thousand years before reaching the true edge of the solar system, the Oort cloud of comets.[4] The Oort cloud is named after the Dutch astronomer Jan Oort, who suggested the concept in 1950. From about 1,000 AU out to 100,000 AU from the Sun, the Oort cloud of

[4] As early as 1993, Voyager science team members announced the first direct evidence that the spacecraft had encountered the so-called heliopause, a region in space where the charged solar wind particles and associated magnetic field streaming radially away from the sun first encountered the charged particles and magnetic field in the interstellar gas. Much was made about the spacecraft then leaving the solar system.

more than one hundred billion icy, comet-like bodies represents the limit of the ability of the Sun's gravity to hang onto these loosely bound icy bodies. The Oort cloud is the source of the so-called long-period comets. Compared to near-Earth asteroids, active long-period comets rarely approach the Earth. However, an Earth-threatening encounter by a long-period comet could be a nightmare scenario since many of them are huge and most would not begin to turn into active, easily discovered comets until they got closer to the warming Sun than Jupiter's orbit. A long-period comet whose orbital aphelion reached the edge of the Oort cloud would take about ten million years to travel the distance from the outer Oort cloud to within the distance of Jupiter from the Sun, but it would then take only nine months to travel the distance between Jupiter's orbit and that of the Earth. Even if one of these long-period comets were discovered just outside the orbit of Jupiter and then found to be on an Earth-threatening trajectory, there would be only a few months to do anything about it (see chapter 10).

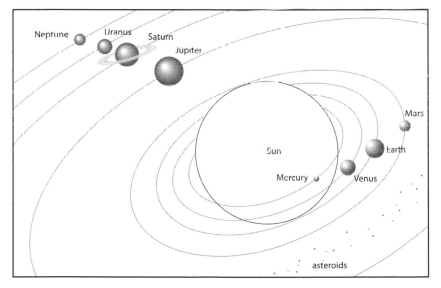

FIGURE 2.2. This schematic illustration shows the rough distances of the eight planets from the Sun. The respective semi-major axes of Mercury, Venus, Earth, Mars, Jupiter, Saturn, Uranus, and Neptune are 0.4, 0.7, 1.0, 1.5, 5.2, 9.5, 19.2, and 30.1 AU.

Comet Hale-Bopp made a memorable 1997 appearance in the inner solar system, and its solid nucleus has been estimated as large as sixty kilometers in diameter. It is now bound in a closed elliptical orbit about our Sun that extends out to about 360 AU and has an orbital period of about 2,400 years, so it spends most of its time in the scattered disk region beyond the Kuiper belt rather than the Oort cloud. It is gravitationally bound to our Sun. In contrast, when the Voyager I spacecraft reaches the solar system's boundary at the outer edge of the Oort cloud, it will escape our solar system entirely. It will be headed toward no particular star but will have traveled more than a third (37 percent) of the distance to the nearest star other than our Sun,

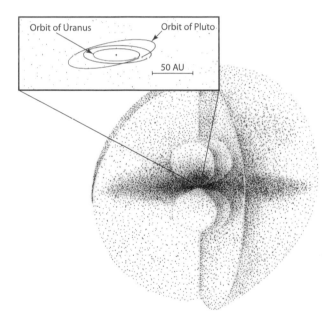

FIGURE 2.3. The Kuiper belt, a relatively flattened distribution of comet-like bodies, is located beyond the orbit of our outermost planet, Neptune, and extends to about 50 AU from the Sun. The dwarf planet Pluto is one of the largest members of the Kuiper belt. The scattered disk of comet-like objects is far less flattened than the Kuiper belt and extends out to 300 AU or more. One hundred billion comets reside in the Oort cloud, the outer edge of which at about 100,000 AU marks the distant boundary of our solar system where the Sun's gravity can no longer hold onto an Oort cloud comet.

Proxima Centauri, which is 4.24 light years distant or approximately 268,000 AU. At its current velocity, Johann and Chuck will reach this distance in 75,000 years or so.

The golden record was affixed to the Voyager I spacecraft with little hope of it ever reaching a pocket of intelligent life on a planet circling a distant star in our Milky Way galaxy. The chances of an alien race enjoying Bach's Brandenburg Concerto or Chuck Berry's rock and roll classic even hundreds of thousands of years in the future are vanishingly small. Even so, we can hope that the spacecraft will one day sail into an inhabited planetary system surrounding a distant star where an advanced civilization may give this tireless, interstellar traveler a home in its museum. It will appear to them as an awkward and crudely built craft, but it will be a testament to the curiosity of a distant race that dared to reach out beyond their own Earthly shores into the unexplored ocean of space.

The Solar System's Current Configuration

Roughly 99.9 percent of the solar system's mass is in the Sun and of what's left, Jupiter's mass is more than one and a half times bigger than all the other solar system bodies combined. So the Sun rules and Jupiter is the big kid on the planetary block. Saturn is the second most massive planet so, not surprisingly, Jupiter and Saturn control the evolution of asteroids from the asteroid belt into the near-Earth neighborhood. Table 2.1 provides the distances from the Sun and the approximate masses of the planets, comets, and asteroids.

The Origin of the Solar System

The measurement of isotope ratios in ancient meteorites suggests the solar system began forming about 4.6 billion years ago.[5] Rewinding

---

[5] Meteorites, including metallic samples from Meteor Crater near Winslow, Arizona, have been age dated using the process of radioactive decay of lead isotopes in the sample. By measuring the concentration of the stable end product of the decay process, together with knowledge of the decay product's half-life and the original concentration of the decaying element, the age of a meteorite can be determined.

TABLE 2.1. Distances from the Sun and relative masses for the planets, Pluto, and the solar system's small bodies

Body	Approximate distance from the Sun (in AU)	Mass (in units of the Earth's mass)
Sun	0	333,000
Mercury	0.39	0.055
Venus	0.72	0.815
Earth	1.0	1.0
Moon	1.0	0.012
Mars	1.52	0.107
Main asteroid belt	2–4	6×10^{-4}
Jupiter	5.20	317.83
Jupiter Trojans	5.2	10^{-5}
Saturn	9.54	95.16
Uranus	19.18	14.54
Neptune	30.06	17.15
Pluto	39.47	0.002
Kuiper belt	35–50	<0.1
Oort cloud	1,000–100,000	4–80

Note: The Sun's mass is 1.99×10^{30} kg and that of the Earth is 5.97×10^{24} kg.

the tape of the solar system's history for that length of time requires a supposed sequence of events that would logically get from a presumed original configuration to what we see today. Scientists often use the term "model" to describe a system of postulates, mathematical relationships, data, and inferences that form a basis for describing a scientific process or sequence of events. These models need to be tested against new observations to see how well they hold up. That's a big part of the scientific method. Do the predictions provided by the model match the new observations? If so, great, and score one for the model. If not, a new model or at least a modification to the old one is required before further predictions can be provided—and tested yet again. Models, such as creationism, that cannot be tested are based upon faith or passionate assertions and fall well outside science. Until recently, the model for the solar system's origin was thought to be well understood and was taught in countless astronomy textbooks as the nebular hypothesis. But then some pesky observations within our solar system and in other extra-solar planetary systems didn't agree with the traditional model. First let's look at the traditional solar system formation model—or nebular hypothesis—and then we'll examine the observations that don't quite fit this traditional model. In the next chapter we'll touch upon a new model that can explain some of the deficiencies of the nebular model for our own solar system formation process. It also has fewer problems explaining some of the peculiar planetary systems that have recently been discovered around other stars. This new model is called the Nice (pronounced "niece") model because it was developed by scientists resident in, or temporarily in, Nice, France— although the model itself is quite nice.

The modern nebular hypothesis posits that our solar system began as an enormous cloud of gas and dust much like many such clouds that are currently observed in several locations within our Milky Way galaxy. This cloud's composition was almost entirely hydrogen, with some helium and dust particles that included most of the elements in the periodic table. Our primordial "protoplanetary" cloud was not uniform in density but had clumps of material throughout. As gravity caused the cloud to collapse, the slowly rotating cloud would have spun faster to conserve angular momentum, the same reason an ice skater twirls faster when she pulls her arms into her body. This increased spin had the effect of counteracting gravity in the spin plane but not in the direction perpendicular to this plane. Thus the cloud of gas and dust evolved into a rotating Frisbee-shaped disk, the central region gained more and more mass, and it became more and more compact. When its temperature reached one million degrees, hydrogen began to burn to form helium, a fusion reaction. The Sun was born. This initial collapse of the protoplanetary nebula is thought to have occurred in less than one hundred thousand years, which is extremely rapid by solar system standards.

On a much longer time scale of ten to one hundred million years, the dust particles then settled to the mid-plane of the protoplanetary disk and collided with other dust particles at low velocities. Because of electrostatic forces, the particles stuck to one another with subsequent low-velocity collisions, creating more compact and larger particles. As long as the relative velocities between agglomerating particles remained low enough, a process of runaway growth could have taken place because the more massive a particle becomes, the more effective it is in gathering to itself all the neighboring particles.[6] These planetary building blocks, or planetesimals, are thought to have reached kilometer- and larger-sized entities before agglomerating into the protoplanetary bodies themselves. The ultimate size that a protoplanetary body could achieve depended upon its distance from the Sun as well as the density and composition of the nebula where it formed. Any massive protoplanetary body that could slowly approach a less massive planetesimal as they both orbited the Sun would have captured it. The nearby planetesimals that were not slowly overtaken by the massive body's orbital motion could have been stirred up to such an extent that their own attempts to form a separate planet would have been frustrated. Jupiter's enormous mass is believed to have come about as a result of a runaway mass growth process; it was capable of stirring up and increasing the relative velocities of the planetesimals interior to it. As a result, the present-day bodies in the asteroid belt were incapable of agglomerating into a planet of their own. Had this not been the case, there would be very few near-Earth asteroids.

Water Worlds

In the protoplanetary disk, the water molecule was the second most abundant molecule (H_2O) after hydrogen (H_2). The element helium would have been far more abundant than oxygen in the disk, but helium is inert and does not combine easily with any other elements. In contrast, oxygen, which is the third most abundant element in the solar system after hydrogen and helium, is far friendlier and combines easily with the most abundant element, hydrogen, to form the water

---

[6] The primary reason why astronomers demoted Pluto to a dwarf planet in 2006 was due to its not meeting one of the recently established criteria for being a planet. Unlike the other planets, Pluto was not capable of completely clearing its orbital zone of bodies of comparable mass. That is, it did not reach a sufficient mass to draw to itself all the nearby planetesimals of comparable size in its orbital path about the sun. Pluto did meet the other two planet membership criteria: it was in orbit around the sun without being some other planet's satellite, and it had sufficient mass to draw itself into a nearly spherical shape.

molecule. About one million years after its formation the Sun is believed to have entered a flare stage, or a so-called T-Tauri phase, whereby the lighter gasses of the protoplanetary nebula were blown out of the inner solar system, leaving the inner planets rocky and devoid of the hydrogen and helium that are so evident in the two outer gas giant planets, Jupiter and Saturn.

The so-called snow line, which is actually a theoretical surface, divides the inner, relatively warm regions of the protoplanetary disk from the cold and ice-rich regions of the outer disk. Inside this line, water existed in the form of vapor while for the colder temperatures beyond the snow line, the water vapor condensed into ice grains that could slowly collide, stick to one another, and allow for rapid growth of the existing planetesimals. For the current temperature distribution within the solar system, the snow line would be near 4 AU, but the increased opacity caused by the dust in the denser initial protoplanetary disk could have moved the snow line closer to Mars. As a result some main-belt asteroids could have formed with water ice, and in fact, a few active icy comets in the main asteroid belt have been recently discovered.[7] Most solar system formation models suggest that the Earth formed dry because of the Sun's heat inside the snow line and the additional heat generated by the planetesimal bombardment process that formed the early Earth. These collisional events formed the Moon and the subsequent so-called Late Heavy Bombardment that rained planetesimals down upon the Earth and Moon about 3.9 billion years ago. The current Earth's extensive oceans then need to be explained; in chapter 4 we'll discuss how near-Earth objects bearing water ice might have supplied at least some of this water. Liquid water is required to sustain complex, carbon-based life as we know it and Earth is the only planet within our solar system that has an appreciable supply of liquid water on its surface—about 0.02 percent of the Earth's mass. Our Earth is located in the Goldilocks habitable zone of our solar

---

[7] There have been at least five: 133P/Elst-Pizarro, 176P/LINEAR, 238P/Read, P/2008 R1 Garrard, and P/2010 R2 La Sagra. In addition, the water ice spectral band center near 3.1 microns has been identified on the surface of asteroid 24 Themis by two teams using the NASA Infrared Telescope facility on Mauna Kea, Hawaii.

system that is just right for our planet to maintain bodies of liquid water without being constantly frozen and yet far enough away from the Sun so that our waters do not boil away. Further, we're fortunate that Earth's water does not completely cover the landmasses since technological civilizations probably require dry land as well as water.

The vast majority of the primitive planetesimals did not survive the solar system formation process. They were either subsumed into forming the planets or they were gravitationally perturbed by the giant planets, especially Jupiter, into a death plunge at the Sun or kicked out of the solar system entirely. However, a tiny percentage of these witnesses to the solar system's birth did survive. In the inner solar system, some rocky planetesimals remained in the asteroid belt while in the outer solar system, some icy planetesimals formed and stayed put in the Kuiper belt. Neptune then gravitationally kicked some of these bodies into the scattered disk region. A large number of these icy bodies that had formed inside the Kuiper belt region between the orbits of Jupiter and Neptune had their semi-major axes, perihelia, and orbital inclinations raised via gravitational interactions with Uranus and Neptune to form the giant Oort cloud.

Our discussion of the solar system formation process suggests there are compelling scientific reasons to study the comets and asteroids that make up the near-Earth object population. In general, the leftover bits and pieces from the inner solar system planetary formation process are what we now call asteroids, and the ancient remnants from the outer planetary formation process are the comets. Asteroids and comets are perhaps the most primitive, least changed remainders from the early solar system formation process. If we'd like to study the chemical mixtures and the thermal environment from which the planets formed some 4.6 billion years ago, then the intense study of asteroid and comet samples is critical. The aptly named near-Earth objects are the easiest to study with ground-based and space-based scientific instruments. They are the obvious target bodies for returning asteroid and comet surface samples to Earth for in-depth studies of their physical nature, their detailed elemental and mineralogical compositions, their ages, and the intense heating environment that some of them have undergone.

Flies in the Ointment

The modern nebular model has an appealing simplicity and logic. The rocky inner planets (Mercury, Venus, Earth, and Mars) are relatively small because most of the protoplanetary dust and gas were blown out of the inner solar system before they formed. Because Jupiter, Saturn, Uranus, and Neptune were beyond the snow line where the water vapor could condense to ice crystals that would stick to one another, these outer planets were either gas giants (Jupiter and Saturn) or ice giants (Uranus and Neptune). Jupiter had runaway growth that frustrated the formation of a planet between the orbits of Mars and Jupiter, leaving the asteroid belt there as a testament to Jupiter's dominance. Farther out beyond Neptune, the protoplanetary nebula would have been less dense as the distance from the Sun increased so the bodies in the Kuiper belt and the comets that were thrown out into the Oort cloud would be far more modest in size. Yes, the modern nebular model for the solar system's origin is appealing. Unfortunately, there are several flies in the ointment; four of them are noted below.

> Nebular models of the solar system formation cannot account for the current masses of Uranus and Neptune if they formed at their current distances from the Sun (i.e., 19 and 30 AU).

> Current models suggest that there would have to be more than 10 Earth masses of material in the Kuiper belt to form objects of the size of Pluto and the comparably sized Kuiper belt object Eris, but there is less than 0.1 Earth mass of material there now.

> Current models predict an Oort-like cloud of comets beyond the planetary system that is far less massive than what is actually present in our solar system.

> Recent observations of planetary systems outside our own solar system do not generally follow the pattern of small rocky bodies in the inner planetary system with ice and gas giants forming beyond the snow lines of these systems.

In the next chapter, we'll investigate the Nice model that removes these flies in the ointment by positing the remarkable suggestion that the giant outer planets and the Kuiper belt did not form where they now reside.

CHAPTER 3

How and Where Do Near-Earth Objects Form?

Comets and asteroids are the leftover bits and pieces from the
early solar system formation process.

The Gravity Assist

Mother Nature is conservative, especially when it comes to the orbital
energies of her solar system bodies. During any encounter between two
bodies in space, energy is conserved in that if one body gains energy
during an encounter, the other one loses an equal amount of energy.
For example, when the Voyager I spacecraft careened past Jupiter in
1979, the spacecraft received a huge boost in its orbital energy and
Jupiter suffered an equal orbital energy loss. Of course since Jupiter is
two trillion trillion times more massive that the spacecraft, it suffered
far less of an effect than a speeding semitrailer truck would feel from
the gravitational effect of a close approach by a housefly. Nevertheless,
the same conservation of orbital energy was key to the initial formation of the solar system when millions of planetesimals made encounters with the far larger planetary bodies.

A planetesimal's orbital energy is proportional to its orbital semi-
major axis. A body whose orbital energy is increased will have its semi-
major axis increased proportionately. Its orbit will expand in size.
Planetesimals orbiting the Sun in the neighborhood of a massive
planet can, from time to time, pass close by this planet and in doing so

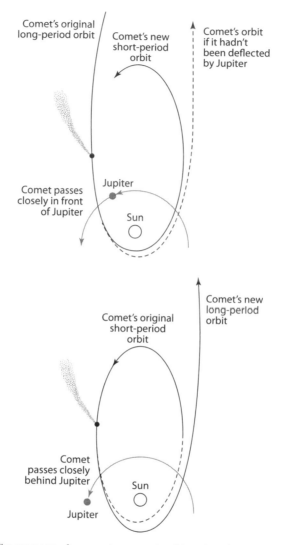

FIGURE 3.1. The passage of a comet or an asteroid past a planet will alter its future orbit, either removing orbital energy (reducing its orbital period) with a front-side passage of a planet or adding orbital energy (increasing its orbital period) with a trailing edge passage. For example, in the top frame, a comet makes a front-side passage of the planet Jupiter, loses orbital energy, and enters into an orbit with a shorter orbital period than before the Jupiter approach. In the bottom frame, the comet passes behind the planet Jupiter, gains orbital energy, and has its orbital period increased as a result. Space mission planners often take advantage of these planetary gravity assists to save rocket fuel when sending spacecraft to distant locations within the solar system.

can either gain orbital energy or lose it with a concomitant increase or decrease in their semi-major axis. As depicted in figure 3.1, a backside passage of the comet planetesimal with respect to the more massive planet would increase the orbital energy and semi-major axis of the planetesimal with an equal loss of energy by the planet. Likewise, a front-side passage of the planet by the comet planetesimal would decrease its orbital energy and semi-major axis while the planet would gain a corresponding amount of orbital energy. However, since the planet is so much more massive than the planetesimal, its change in semi-major axis is negligible for a single encounter. But over hundreds of millions of years of encounters, these tiny changes in a planet's semi-major axis can add up to a significant amount.

The Uruguayan astronomer Julio Fernández and his Taiwanese colleague Wing Ip first suggested planetary migration during the solar system formation process in a 1984 publication. It would prove to be a seminal work although it did not receive a great deal of attention at the time. Eleven years later, the Arizona scientist Renu Malhotra effectively used the planet migration concept to explain Pluto's unusually high orbital eccentricity (0.25) and orbital inclination (17 degrees). More recently, planetary migration has proven to be key to modern solar system formation theories because the concept can remove many of the problems with the nebular solar system formation outlined in chapter 2. However, there still remain one or two flies in the ointment.

Planetary migration during the early solar system formation process seems unavoidable. It is a back reaction on the planets themselves of the planetesimal scattering process that populated the Oort cloud. A single massive planet scattering a population of planetesimals in near-circular orbits in the vicinity of its own orbit would suffer no net change in orbital radius since it scatters approximately equal numbers of planetesimals inward and outward. However, in the real solar system, the subsequent fate of these planetesimals is not symmetrical.

Let's consider the fates of Neptune and Jupiter during the solar system formation process. Both planets were formed from neighboring icy planetesimals that were gravitationally agglomerated into the larger planets. Once formed, both planets and the remaining belt of leftover planetesimals beyond Neptune moved about the Sun in nearly circular orbits. Frequent gravitational interactions between Neptune and the planetesimals in the inner belt region continued to occur with some additional planetesimals being drawn into the planet. But most encounters simply scattered them in different directions. At first, Neptune scattered roughly an equal number of planetesimals inward toward the Sun and outward. Most of the icy bodies thrown outward by Neptune either populated the Oort cloud of comets or returned to the outer planetary system near the orbit of Neptune. But Neptune lacked the ability to toss many of these bodies completely out of the solar system. Uranus, then Saturn, and finally Jupiter affected the planetesimals that Neptune scattered inward. Compared to Jupiter, Neptune is a bit of a planetary wimp since it is nearly twenty times less massive. Uranus and Saturn, also a bit wimpy compared to Jupiter, scattered planetesimals inward and outward with few of the latter actually being thrown completely out of the solar system. Once the inward-directed planetesimals reached Jupiter's realm, Jupiter had no problem ejecting them from the solar system. The upshot of this celestial pool game was that Jupiter was very effective in removing the lower-energy planetesimals scattered inward by Neptune so the planetesimal population encountered by Neptune at later times was increasingly a group whose orbital energy was higher than that of Neptune itself. Subsequent scattering events then tended to increase the orbital energy and semi-major axis of Neptune. To a lesser extent, the semi-major axes of Uranus and Saturn also increased, but the orbit of big bully Jupiter decreased a bit since it had to lose orbital energy to compensate for all the energy boosts it gave to the planetesimals that it tossed completely out of the solar system.

The Solar System's Formation: A Nice Model

In the primordial solar nebula, the dust grains and ices agglomerated into countless planetesimals. Almost all of these building blocks of the planets were then consumed by planet formation, were ejected from the solar system entirely, or fell into the Sun. A relatively tiny fraction survived, but even so this remaining population of millions of planetesimals was instrumental in sculpting the solar system that we see today. Four scientists have developed the so-called Nice model that seems to explain much of this sculpting process. Alessandro Morbidelli, Hal Levison, Kleomenis Tsiganis, and Rodney Gomes spent much of their research time in Nice on the French Riviera overlooking the Mediterranean Sea. They really toughed it out. The model they developed became known as the Nice model or, as they slyly refer to it, a Nice model.

The model is a computer simulation that begins with several thousand test particles or simulated planetesimals that are assigned initial positions and velocities and then are allowed to gravitationally interact with the computer-simulated planets to see what happens over millions of simulated years. The goal was to keep adjusting the initial masses and positions of the planetesimals and the initial positions of the protoplanets, let them gravitationally interact with one another, and ultimately wind up with a solar system computer model that closely resembled the way the solar system appears today. It is very labor-intensive work, but of course the computer is doing most of the work while the four scientists sit and admire the view of the Mediterranean Sea and think about their next computer simulation. This probably isn't true; it's just that most of us scientists in the field are a bit jealous because we don't have these views from our offices and cubicles.

Mercury, Venus, Earth, Mars, Jupiter, and Saturn are believed to have formed about the same time 4.6 billion years ago. Uranus and Neptune took longer and reached maturity a few hundred million years later. The Nice computer model begins with the major planets already formed and assumes the primordial, circumsolar gaseous nebula has already been blown out of the solar system, leaving behind a solar system composed of the Sun, planets, and a leftover debris disk of smaller planetesimals. One successful model began with Jupiter, Saturn, Uranus, and Neptune being born on circular, coplanar orbits at distances from the Sun of 5.45, 8.18, 11.5, and 14.2 AU, respectively. These distances can be compared to these planets' current distances from the Sun: 5.2, 9.5, 19.2, and 30.1 AU. Within the Nice model, a dense region of planetesimals, with a total mass of 35 Earth masses, was located in a flattened torus that spanned the region from 15.5 to 34 AU from the Sun—just outside the initial planetary region.

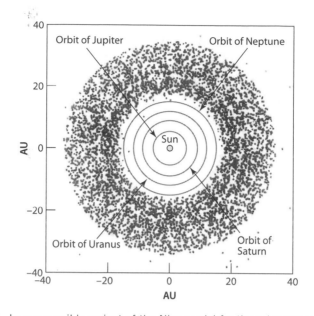

FIGURE 3.2. In one possible variant of the Nice model for the solar system's origin, the outer planets Jupiter, Saturn, Uranus, and Neptune began not on their current orbits but on nearly circular orbits at respective heliocentric distances of 5.5, 8.2, 11.5, and 14.2 AU. The model assumes that there was an initial dense region of planetesimals just outside the distance of the outermost planet, Neptune, which extended from about 15.5 to 34 AU. Over hundreds of millions of years as a result of interactions between the planets with one another and with the planetesimals, Jupiter moved a bit sunward to its current orbital position while Saturn, Uranus, and Neptune moved outward to their current orbits. In the planetary migration process, 99 percent of the original planetesimals were scattered away from their initial orbital locations.

At first, the four major planets lazily orbited the Sun, scattering nearby planetesimals that had been passed down to them by the nearest giant planet farther out. They underwent a slow migration process. Jupiter moved toward the Sun just a bit while the other three planets, especially Neptune, spiraled out from the Sun. The Nice model then produces a triggering event that dramatically disrupts this orderly planetary migration process. One possibility occurs after several hundred million years of planetary migration when Saturn reaches a heliocentric distance of 8.65 AU with a corresponding orbital period about the

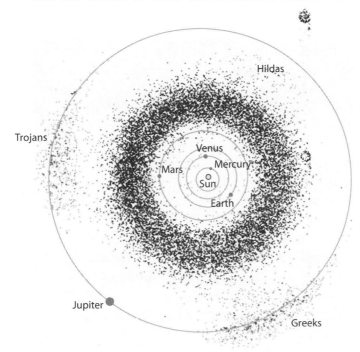

FIGURE 3.3. This current configuration diagram for the inner solar system shows the main belt of asteroids between the orbits of Mars and Jupiter, the so-called Trojan asteroids that lead ("Greeks") and follow ("Trojans") Jupiter by an average of 60 degrees and the Hilda-class of asteroids that have orbital periods two-thirds that of Jupiter. At their aphelia, the Hildas are either on the other side of the Sun from Jupiter's position or near, but slightly inside, the Trojan asteroid locations so they, like the Trojans themselves, avoid the strong gravitational perturbations by Jupiter.

Sun of 25.4 years. Jupiter, near 5.45 AU from the Sun, then had an orbital period just one half that of Saturn so every two orbits of Jupiter would put these two giants in the same relative position with respect to one another. Every 25.4 years they would gravitationally tug on one another, thus increasing their respective orbital eccentricities to values comparable to their current values. Scientists refer to this as Saturn and Jupiter entering into a two to one (2:1) mean motion resonance because the orbital periods of Saturn and Jupiter are then in a ratio of 2:1. The orbital eccentricities of Uranus and Neptune were also increased due to the gravitational tugs of Jupiter and Saturn.

As a result of their initial orbit proximities and the increase in their eccentricities, the orbits of the major planets could now cross one another and planet-planet close approaches became possible. That was a recipe for subsequent chaotic motion. As a result of the migration process and this chaotic motion, the orbits of both the ice giants Uranus and Neptune kept expanding and they ultimately plowed into the planetesimal disk that was located just outside the original orbit of Neptune. The resulting scattering of these planetesimals by Uranus and Neptune then increased dramatically with more and more planetesimals being directed inward toward Saturn and Jupiter. Thus, the planetary migration process increased abruptly and did not cease until the disk was almost depleted. The orbit of Neptune migrated outward until it reached 30 AU from the Sun and then stopped because the original planetesimal disk ended just outside this distance. With no more objects to scatter, Neptune could no longer migrate outward.

This accelerated planetary migration process within the Nice model produces a number of features and peculiarities of the present-day solar system configuration.

The formation of the outer giant planets is now possible since they formed closer to the Sun than their current positions would indicate and hence had a more dense population of planetesimals from which to form. The Nice model indicates that if the initial disk of planetesimals was 35 Earth masses, the rates and distances that the giant planets would have migrated is just enough to explain their current locations within the solar system. In addition, the deviation of the current giant planet paths from circular, coplanar orbits is easily explained.[1]

According to the Nice model, the so-called Trojan asteroids of Jupiter were formed in orbits that have relatively high inclinations up to about 40 degrees, which agrees with the current observed population of Trojans. These Trojans travel in Jupiter-like orbits that either lead or trail Jupiter itself by an average of 60 degrees. As a result, they do not approach Jupiter but fly in formation with it at the same distance from the Sun. As a result of Saturn and Jupiter entering into the 2:1

---

[1] In the planetary migration process, the orbits of Uranus and Neptune can cross one another, and in about 50 percent of the computer simulations Neptune can actually begin with an orbit interior to Uranus then leapfrog to its final orbital position outside the orbit of Uranus.

resonance, groups of scattered planetesimals became chaotic, their inclinations would have increased, and after this period of instability ended some bodies were trapped into the Trojan orbits. In the traditional nebular hypothesis, the Trojan population would have formed in much lower inclination orbits. The same 35 Earth mass disk of planetesimals that was required to get the correct planetary migration rates also leads to a total mass of the Jupiter Trojans equal to 0.000013 Earth masses in the Nice model, a figure that is comparable to the estimated mass of the current Trojan clouds. Although these bodies are most often referred to as Trojan asteroids, they are likely to be inactive icy bodies.[2]

Analogous to the formation of the Trojan asteroids, the chaotic planetesimal motions induced by the 2:1 resonance can also explain the presence of some of the irregular satellites of the giant planets— satellites that can be in distant, inclined, or even retrograde motions about their primary planets. Once the instability of the 2:1 resonance subsided, the giant planets could have captured many of their outer irregular satellites.

The Nice model naturally forms the Kuiper belt planetesimals that extend from about 35 to 50 AU from the Sun. Although about 99 percent of the planetesimal disk mass was scattered away during the planetary migration process, a sizable number remained just beyond the current orbit of Neptune. These objects reside in the Kuiper belt and since the disk was once about one hundred times more massive than it is now and closer to the Sun, the existence of relatively large Kuiper belt objects like Pluto, Eris, and Makemake is understandable.

The scattered disk objects that currently reside beyond the Kuiper belt can also be explained as a result of the planetesimal scattering process. These objects, containing frozen volatiles such as water and methane, were scattered outward by Neptune and now have their perihelia near the inner edge of the Kuiper belt while their aphelia can extend to well beyond 300 AU. Because they can still be scattered by Neptune, they are on unstable orbits and are the likely source for the centaurs and the short-period comets. That is, Neptune can perturb the scattered disk objects and send them first into the region between

---

[2] In addition to the Jupiter Trojans, Neptune has several Trojan-like objects flying in formation with it about the sun. Mars also has a few Trojan-like objects and the Earth has one (2010 TK7).

Jupiter's orbit and that of Neptune where these objects are called centaurs.[3] Subsequent planetary gravitational tugs can then push these centaur objects closer to the Sun and into Jupiter's neighborhood. Jupiter can either throw them out of the solar system or send them into the inner solar system where the warming Sun can vaporize their ices, thus releasing the gas and dust that converts them from icy planetesimals into active comets.

The roughly spherical Oort cloud of comets exists at distances between about 1,000 AU out to the very edge of the solar system at 100,000 AU. It is a natural by-product of the planetary migration process whereby Neptune and Saturn scattered icy planetesimals outward but not usually with enough energy for these bodies to completely escape the solar system. Once these icy planetesimals are in the Oort cloud, passing stars or the tidal effects of the Milky Way plane of stars can perturb these loosely bound comets into the inner solar system where occasionally they can be seen as the visually impressive long-period comets. However, whereas the Oort cloud is believed to have somewhere between 4 and 80 Earth masses of material spread among about 400 billion comets, the Nice model can only produce less than 2 Earth masses of material in the Oort cloud. The Nice model that was batting nearly a thousand in successfully predicting the current state of the solar system sort of drops the ball with regard to the mass of the Oort cloud. Either the Nice model is lacking is some detail or possibly some of the Oort cloud comets were captured from another star while the solar system was forming within a star cluster.[4]

The rain of impacting material suffered by the Moon and Earth about 3.9 billion years ago, the Late Heavy Bombardment, can also be attributed to the instability introduced into the solar system when Saturn and Jupiter reached the 2:1 resonance. The chaotic motion introduced by this resonance passage stirred up the outer giant planets and sent Neptune plowing into the disk of planetesimals. As a result, many of these planetesimals would have been pushed into the inner solar system

[3] In 1977 Charles Kowal discovered the first centaur, (2060) Chiron, between the orbits of Saturn and Uranus.

[4] In 2010 Hal Levison and his colleagues suggested that 90 percent of the Oort cloud could have been captured from other stars while the Sun was in its birth cluster ("Capture of the Sun's Oort Cloud from Its Birth Cluster," Science, June 10, 2010).

where they would be expected to hit the Sun, be thrown out of the solar system by Jupiter, or hit planets interior to the planetesimal disk. The Earth and Moon would not be immune from this inundation of icy outer solar system material. Members of the outer asteroid belt nearest Jupiter would also be expected to have their orbits significantly modified by the chaotic behavior induced by the 2:1 resonance crossing and they, too, would contribute to the Late Heavy Bombardment of the Earth and Moon. Almost all of the ancient craters on Earth have been cleared away by tectonic evolution as well as by wind and water erosion, but the Moon's cratering record is still available for study. There are more than 40 lunar basins larger than 300 kilometers and all of them are thought to be older than 3.8 billion years. The best age determinations are available for the large lunar basins Imbrium and Orientale, which are, respectively, 3.85 and 3.82 billion years old. Clearly there was a dramatic decrease in impacting planetesimals after about 3.9 billion years ago.[5]

Relatively recent observations of several hundred extra-solar planetary systems have been made, often using the indirect technique of observing the wobble of a particular star and using the periodicities and magnitudes of these wobbles to infer the masses and distances of their orbiting planetary bodies. Many of these planetary systems have giant planets orbiting very close to the parent star, which flies in the face of solar system formation models that attribute rapid planetary size growth to their being located outside the snow line where ice particles can form and stick to one another. However, these close-orbiting "hot Jupiters" could have formed farther out beyond their snow lines and then migrated inward to where they now reside. However, we must bear in mind that the easiest planets to discover would be the most massive ones that are relatively close to their parent stars and thus capable of introducing the most obvious wobbles. Hence there is an

[5] During the Late Heavy Bombardment the estimated total mass that impacted the moon was roughly 10^{22} grams, which would be equivalent to a single rocky impactor with a diameter of 194 kilometers. A larger quantity of material would have hit Mars, and the cometary portion of this influx may have provided the ices that still exist on the Martian surface and subsurface. Likewise, the lunar ices may also have been delivered by a cometary influx. The Late Heavy Bombardment theory was put forward as a result of the age dating of the Apollo lunar samples although Ralph Baldwin, in his classic 1949 work *The Face of the Moon*, noted that there must have been "a very rapid accumulation of meteoric material early in the moon's career."

observational bias in favor of finding large and close planetary bodies around other stars. Even so, extra-solar planet-finding techniques are becoming more and more sophisticated and there are several discoveries of "super Earths" that are up to ten times more massive than the Earth but still quite possibly in the habitable Goldilocks zone where life could form in the presence of liquid water. Recent estimates suggest that a few percent of all Sun-like stars have Earth-sized planets in orbit about them. Even if only 1 percent of the stable Sun-like stars in our galaxy have Earth-sized planets in their habitable zones, that is an enormous number of possible life-bearing planets among the one hundred billion stars in our Milky Way galaxy—and then there are the hundred billion other galaxies.

The Near-Earth Object Population

The vast majority of near-Earth objects are inactive asteroids born in the asteroid belt between the orbits of Mars and Jupiter. Only a very few, about 1 percent of the total near-Earth population, are active short-period comets that arrive in the Earth's neighborhood after having been scattered inward by the giant planets. According to the Nice model, the asteroids that currently reside in the outer regions of the asteroid belt are primitive, relatively unaltered bodies that were affected by the chaotic motions within the early planetary system. Most of these bodies in the outer asteroid belt are very dark, blacker than coal objects that contain primitive carbon-based materials. Some of them likely contain surface or subsurface ices or hydrated minerals that contain bound water or hydroxyl (OH) radicals. Although the observational evidence is not very clear, the Nice model would suggest that these outer main-belt asteroids should be similar in composition to the Trojan asteroids, the centaurs, and the Kuiper belt objects. If some of these dark outer main-belt asteroids were to venture into the inner solar system, the Sun's warming rays might activate any ices present, whereupon they would be called comets since an icy planetesimal can be a comet if it's active or an asteroid if it's not. The line between comets and some asteroids is not clearly drawn.

Most near-Earth asteroids were liberated when larger asteroids had catastrophic collisions in the inner asteroid belt. According to the Nice model, they did not participate in the chaotic motions that sculpted the planetary system farther out. The majority of these bodies would then have spent their lifetimes relatively close to the Sun so most would be expected to be devoid of ices. However, there are some very dark, inner asteroid belt bodies that probably drifted inward from the outer asteroid belt and they could harbor hydrated minerals, perhaps even deep subsurface ices.

Scientists seek order and guiding principles whereas Mother Nature sometimes seems to do her best to thwart these attempts. So it is with the attempts of astronomers to place various asteroids into classes and subclasses by using telescopic observations of their reflectivities in various visible and near-infrared wavelengths of light. These spectral characteristics have been used to classify many asteroids into the relatively bright S-class (silicate rock including olivine and pyroxene), the much darker C-class (carbon-based materials often with hydrated minerals), the M-class (sometimes metallic, sometimes not), and the dark D-class (possibly ex-comets that have run out of gas), along with an alphabet soup of other asteroid classes and subclasses that may provide hints as to the minerals and materials that make up these bodies. However, the only surefire technique to discern Mother Nature's design and positively identify the elements and minerals in a particular asteroid is to send a space mission there, bring back a sample, and study this sample with a variety of complex ground-based instruments.

Sources and Sinks for Near-Earth Objects

Although active short-period comets are part of the near-Earth object population, their source regions and end states, or sinks, are often very different than those of the asteroids that make up the vast majority of the population. After being perturbed by a passing star or the massive collection of stars in the plane of the Milky Way galaxy, long-period comets often arrive in the inner solar system from the Oort cloud. They can arrive on nearly parabolic orbits after trip times of millions of years. Most short-period comets begin in the Kuiper belt or scattered disk region, and after being perturbed inward by the giant outer planets they eventually fall under the influence of Jupiter, which then controls their subsequent motions in the inner solar system. Most active comets are on unstable orbits, and their dynamic lifetimes are limited to about a million years before they collide with the planets or the Sun or are ejected back to the outer planetary system by one of the giant planets. Comets appear to be far more fragile than asteroids; several have been observed to split into fragments that subsequently dispersed into clouds of dust. While in the inner solar system, active comets can use up all their ices or develop an inactive crust of material that can close off the Sun's access to these ices. So active comets can become inactive or dormant icy bodies that are indistinguishable from a dark D-type asteroid. A few percent of all near-Earth asteroids are thought to be inactive comets. For many comets, their active lifetimes are far shorter than their dynamic lifetimes. They dry up, split, fragment, or dissipate into clouds of dust and gas long before they run into something big or get rudely ejected by Jupiter to the frozen gulag of the outer planetary region or interstellar space.[6]

The source region for almost all near-Earth asteroids is the inner asteroid belt located between 2 and 3 AU from the Sun. If an asteroid is located at select distances in the inner asteroid belt, the gravitational tugs of Jupiter and Saturn can change its orbit over time so that its orbit crosses that of Mars and then reaches the orbit of Earth. For example, an asteroid orbiting the Sun at a distance of 2.5 AU will be in a 3:1 resonance with Jupiter. That is, the asteroid will circle the Sun three times every time Jupiter circles the Sun once and they will arrive at the same place relative to one another every twelve years. Over a period of about a million years, Jupiter can then pump up the eccentricity of the asteroid's orbit, sending it into an orbit that crosses Mars and eventually it can become a near-Earth asteroid. Likewise, the cumulative gravitational effects of Saturn can affect an asteroid in a circular orbit about the Sun near the inner edge of the asteroid belt,

[6] Because of their vaporizing ices and the subsequent loss of gas, dust, and debris clumps, even comets that are not disrupting leave trails of dust and debris in their paths. When Earth sweeps by these dust trails, meteor showers can occur.

near 2.1 AU, and its orbit will be modified into that of a near-Earth asteroid in about a million years. These asteroids then spend an average of several million years in near-Earth space before they disappear into a sink. The most frequent sink for these asteroids is collision with the Sun, followed by Jupiter ejecting them from the solar system, followed by a collision with a planet. But over time, these resonance regions will be completely cleared out as more and more asteroids are sent into near-Earth space and lost to collisions and solar system ejection. Eventually all near-Earth asteroids should vanish into these sinks, but today there's still an enormous number of them so there has to be a source mechanism to push asteroids into these resonance regions so that they, too, can become near-Earth objects. One of the most promising mechanisms for pushing asteroids into these resonance regions is the thermal re-emission of sunlight, often called the Yarkovsky effect after Ivan Yarkovsky, a Russian civil engineer who suggested this phenomenon in 1901.[7]

Yarkovsky and YORP

In the inner solar system, an asteroid will absorb energy from the Sun and re-radiate it into space as heat. If the asteroid were not rotating, the incident and re-radiated energy would occur along the same Sun-asteroid line. But because all asteroids are rotating, the re-radiated energy will emerge from a direction that differs from that of the Sun and the asteroid will emit more heat on its "afternoon" side. The same phenomenon is responsible for making the Earth's afternoon 2:00 hour warmer than its morning 10:00 hour even though a specific location on Earth will receive the same amount of sunlight at each time. Infrared photons departing the afternoon side of the asteroid create a feeble rocket effect that imparts a push. Depending upon whether the asteroid is rotating in the same "prograde" direction that it orbits the Sun or in the opposite

[7] In 1901 Yarkovsky proposed a thermal force pushing on the planets to explain why these bodies were not observed to lose energy as they encountered resistive drag forces while moving through the interplanetary ether. Since we now know there is no interplanetary ether, Yarkovsky gets credit for correctly identifying the effect that carries his name, but his suggestion was made for the wrong reason.

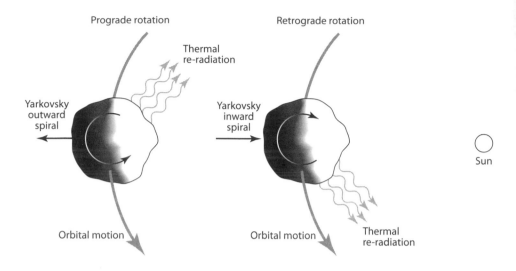

Prograde rotation

Retrograde rotation

Thermal
re-radiation

Yarkovsky
outward
spiral

Yarkovsky
inward
spiral

Sun

Orbital motion

Orbital motion

Thermal
re-radiation

FIGURE 3.4. The so-called Yarkovsky force affects the long-term motion of relatively small asteroids due to a time lag between when sunlight is received upon the surface of the asteroid and when it is re-radiated. Because the asteroid is rotating, there is a small but non-negligible thrust in a direction somewhat different than the Sun's direction. Hence there is a tiny push on the asteroid either in the direction of its orbital motion if the rotation is in the same sense as its motion about the Sun (prograde or direct rotation) or counter to its orbital motion if its rotation is in the opposite sense (retrograde). As shown in the illustration, a prograde rotation will result in energy being added to the asteroid's orbit with a subsequent outward spiral and an increase in the orbital period. A retrograde rotation will subtract orbital energy and decrease the orbital period.

"retrograde" direction, it will gain or lose orbital energy. In other words it will spiral either outward or inward. In 2000, a group of astronomers led by David Vokrouhlický predicted that a well-observed, half-kilometer-sized asteroid named (6489) Golevka would exhibit the Yarkovsky effect when it became observable again in 2003. Three years later when the 2003 observations became available, JPL dynamicist Steve Chesley and his colleagues did need to include a small Yarkovsky effect in their dynamical computer model before they could get it to successfully predict where the asteroid was actually observed to be within the 1991–2003 interval. The rocket-like thrust force was tiny, but because the asteroid had been extensively observed with both optical and radar measurements, the roughly fifteen-kilometer deviation introduced by the

FIGURE 3.5. The YORP effect. As a result of uneven re-radiation of solar energy from one side of a rotating asteroid with respect to the other side, the re-radiation of thermal energy to space can act to increase or decrease the asteroid's rotation rate.

Yarkovsky effect was noticeable and had to be taken into account when predicting the subsequent motion of this asteroid. Steve Chesley had measured a force of about one ounce, or 28 grams, acting upon an asteroid that had a diameter longer than five football fields and weighed 460 billion pounds or 210 billion kilograms. The Yarkovsky force is tiny, but acting over a million years, asteroids can be moved out to, or into, resonance regions where they can then be pushed into the near-Earth neighborhood by the gravitational nudges of Jupiter and Saturn.

If one side of an asteroid has a different shape or reflectivity than the other, one side will have greater thermal re-radiation than the other. This will cause the asteroid to increase, or decrease, its rotation rate. This effect is named after Yarkovsky, John O'Keefe, V. V. Radzievskii, and Stephen Paddack. But since the name "Yarkovsky-O'Keefe-Radzievskii-Paddack" doesn't exactly roll off the tongue, scientists took the first letter from each name and now refer to this phenomenon as the YORP effect.[8]

---

[8] The YORP effect was first detected in 2007 by the Northern Ireland astronomer Stephen Lowry and his colleagues by carefully monitoring the increasing rotation rate of asteroid (54509) 2000 PH5, now designated (54509) YORP.

The YORP effect can increase the spin rate of a relatively small aster-oid to such an extent that if it has a fragile, rubble pile structure that is held together with little more than its own self-gravity then material can fly off its equatorial region and this debris can then re-agglomerate to form a separate satellite body around the primary. Alternatively, the asteroid spin-up can cause it to fission into two components. Roughly 15 percent of all near-Earth asteroids are binary objects and there are at least two triple systems. Current thinking invokes the YORP effect to explain the satellites of most small near-Earth asteroids. A viable binary-producing mechanism for the larger main-belt asteroids involves the collision of two asteroids with two fragments flying off in the same direction and at the same speed, forming a binary pair. Over long peri-ods of time, the YORP effect can cause the rotation axis of a relatively small asteroid to preferentially evolve toward a position that is perpen-dicular to its orbit plane, which is then optimal for the Yarkovsky effect to increase or decrease its semi-major axis.

The Yarkovsky effect slowly transports inner main-belt asteroids into resonance regions where the gravitational tugs of Jupiter and Sat-urn then pump up their orbital eccentricities and turns them into near-Earth objects. Their orbits can then intersect that of the Earth, allowing impacts if both a near-Earth object and the Earth arrive at the same orbital position at the same time. Because the early solar system had far more planetesimals than now exist, the young Earth was bom-barded with planetesimals. One of these collisions likely formed the Moon, then others caused the Late Heavy Bombardment and the lay-ing down of a veneer of carbon-based material and water on the Earth's surface, thus providing the building blocks of life. Once life formed, subsequent collisions then punctuated evolution, allowing only the most adaptable species to evolve further.

CHAPTER 4

Near-Earth Objects as the Enablers and Destroyers of Life

*The question is not whether an asteroid has Earth's
name on it but rather which one and when?*

"Hello, Earth, Mother Nature calling, you haven't been paying attention"

The evidence for impact cratering events within the solar system is obvious and ubiquitous. The first telescopic views of the moon by Galileo in 1609 immediately showed craters, and almost every solid planet or natural satellite whose surface has been observed with high enough resolution shows unmistakable evidence of impact craters. Yes, tectonic evolution, wind and water erosion on Earth, and volcanism and erosion by wind-driven dust on Mars erased many of these craters, but just how clueless could some astronomers be? How many shots across the bow did Mother Nature need to fire before astronomers finally noticed that the Earth runs its course around the Sun in an astronomical shooting gallery—with us as the target?

The written history of impacting near-Earth objects is a short one because the idea of "rocks from space" was not generally accepted until the nineteenth century and the extent of the near-Earth object population was not apparent until the 1990s.[1] Impact events on the Earth and Moon were not generally recognized as such until well into the second half of the twentieth century; the Moon's craters were originally thought to be due to volcanic structures. Those who argued for impact craters on the Moon noted that they differed in shape from terrestrial volcanoes and that some were far larger than any volcanic structures on Earth. But they had to explain why almost all of these craters were circular and yet most impactors in heliocentric orbit would have oblique approach paths and hence would be expected to form elongated craters. With characteristic genius, the Estonian-born astronomer Ernst Öpik pointed out in 1916 that because of the velocities involved, lunar impacts would be explosive and produce nearly circular craters no matter what their angle of approach. Unfortunately, Öpik's correct explanation was published in a nearly inaccessible Russian journal and hence did not have the immediate influence it merited.[2]

In 1893 the eminent American geologist Grove K. Gilbert argued that the lunar craters were formed by impacts of Earth-orbiting satellites but believed that what is now called Meteor Crater near Winslow, Arizona, was formed as a result of a volcanic steam explosion. He based this conclusion on the circular nature of the crater rim, the lack of a magnetic anomaly from a presumed buried iron impactor mass, and the fact that the amount of material in the crater's rim was roughly equal to the amount of material excavated from the crater itself. None of

---

[1] During the morning hours of December 14, 1807, a huge fireball event was witnessed over New England with fragments crashing to the Earth near Weston, Connecticut. A Yale College chemistry professor, Benjamin Silliman, and the college librarian, James Kingsley, collected a number of the meteorite samples, but when these rocks from the sky were brought to the attention of President Thomas Jefferson, he seemed skeptical. Even so, the comment often attributed to him is likely apocryphal: "It is easier to believe that two Yankee professors would lie than that stones would fall from heaven."

[2] Born in Estonia in 1893, Ernst Öpik was only twenty-three at the time of this publication. It was Öpik who first drew attention to another obscure pamphlet that first suggested the Yarkovsky effect. His seven-decade career in astronomy covered an astonishing range of interests and included research contributions in many fields including stellar structure, meteor physics, comets, asteroids, the Earth, Moon, space exploration, cosmology, and astrobiology. He was also an accomplished music composer. I had the privilege to know him in the early 1970s when he would take time from his position as the editor of the *Irish Astronomical Journal* at Armagh Observatory in Northern Ireland to teach summer classes at the University of Maryland, where I was a very impressed graduate student. During the period 1997–2010, Öpik's grandson, Lembit Öpik, served in the British Parliament where he was a champion for near-Earth object studies.

FIGURE 4.1. Meteor Crater, near Winslow, Arizona, is now thought to have been created by the impact of a 40- to 50-meter-sized asteroid of iron approximately 50,000 years ago. The crater is 1.2 kilometers in diameter and 170 meters deep. Although there apparently is no iron mass buried below the crater floor, numerous iron meteorites have been discovered in the region surrounding the crater.
Source: Courtesy of Shane Torgerson.

these observations agreed with what he expected for an impact crater, and his conclusion was accepted by a majority of contemporary geologists who considered this the last word on the subject. An impact origin for the Arizona crater was denied, ignored, and ridiculed by geologists for nearly four decades thereafter. An exception was Daniel Barringer, a lawyer, geologist, and business entrepreneur who insisted that this feature resulted from the impact by a large and commercially valuable iron meteorite. The equipment he used in an unsuccessful attempt to locate the iron mass can still be seen today at the bottom of the crater. The impact was explosive, so apart from small iron meteorite fragments found in the area surrounding the crater, there is no large iron mass beneath the crater. A definitive identification that Meteor Crater resulted from an explosive impact came in 1963 when Gene Shoemaker pointed out the similarities between the physical characteristics of Meteor Crater and those formed by underground nuclear explosions in Yucca Flat, Nevada. He suggested that Meteor Crater resulted from an iron impactor, twenty-five meters in diameter, that struck the Earth with a velocity of fifteen kilometers per second.

Once the importance of impacts by near-Earth objects was finally recognized, a number of early solar system evolution issues became clearer.

Impacts Shape the Moon

The consensus view is that about fifty million years after the Earth's formation, the Moon was created as a result of a Mars-sized impactor that blasted off material into Earth's orbit, which then accreted into the Moon. This lunar formation mechanism, whereby the lunar material is thought to have originated with the impactor's rocky mantle material, most easily explains the lack of a substantial iron core for the Moon, and computer simulations of the impact now can easily explain how such a relatively massive Moon could closely orbit the Earth. The immense amount of energy involved in such a collision could explain the relative depletion of volatiles, like water, on the Moon and its initial molten magma surface.[3] As a result of the collision, the Earth's surface would have been molten and there would have been a transient atmosphere of silicates in the gaseous state from the vaporized rocky mantle. It would have been a hellish environment on Earth with no surface water, no organic carbon-based molecules, and no atmospheric oxygen. It would have been unsuitable for any type of life.

Between the times of the impact that formed the Moon 4.5 billion years ago and the end of the Late Heavy Bombardment 3.9 billion years ago, it is conceivable that very primitive, single-celled life (e.g., bacteria) could have formed. Water and carbon-based materials, two of the building blocks of life, could have been delivered to the early Earth

[3] Alternative lunar formation processes include fission from a rapidly rotating proto-Earth, capture by the proto-Earth, and formation in place at the same time the Earth formed. All of these alternatives have dynamical problems, particularly the capture hypothesis, that requires some sort of unlikely third-body interactions to effect the capture. For the fission alternative, it is difficult to understand how the Earth's rotation can be spun up enough to allow a lunar fission and then spun down to where it is now. In addition, the fission hypothesis cannot explain the relative depletion of volatiles on the Moon. The binary accretion model cannot explain the lack of a substantial iron core for the Moon or its primordial magma ocean.

via near-Earth comet and asteroid collisions. Very primitive life could have begun that early and possibly survived the subsequent hellish environment produced by the Late Heavy Bombardment. It's not altogether clear just when primitive life formed on Earth.

Delivery of Life's Building Blocks to Earth

Complex, intelligent life is a relatively recent development on Earth, but there is fossil evidence for single-celled life as early as 3.5 billion years ago and possibly earlier. These life forms consisted mostly of single-celled, non-nucleated organisms called prokaryotes. The most common examples, still doing very well after 3.5 billion years, are bacteria.[4] Since the hellish environment during the Late Heavy Bombardment would likely have boiled away any existing oceans, there was then a relatively short time in which two necessary building blocks of life, water and carbon-based organic molecules, could have been delivered in quantity to the Earth's surface and then for self-replicating life to form and prosper.[5] So where did these building blocks come from?

The Earth's earliest primordial atmosphere could have been hydrogen rich and was capable, in the presence of an energy source, of synthesizing organics from inorganic compounds.[6] While a hydrogen

---

[4] We humans have ten times as many bacterial cells in and on our bodies than we have human cells. They protect us, aid digestion, and perform far more good than we give them credit for. So suppress that "eew" response and learn to love them.

[5] It is possible that life arrived, pre-formed, on Earth as a result of impacts of dust, comets, or asteroids that carried simple life forms from other solar system bodies or perhaps from bodies that formed outside our solar system. This notion of panspermia has received a good deal of attention, but most scientists reject this possibility since extended periods in space with no protection from cosmic rays and ultraviolet radiation would prove fatal to almost all life forms. The validity of this delivery system has not been decided definitively but, in any case, the arrival of primitive life via panspermia merely pushes the question of life's formation from Earth to another body.

[6] In 1952, Stanley Miller and Harold Urey conducted an extraordinary experiment in which they hoped to test the idea that conditions on the primitive Earth favored chemical reactions that synthesized organic compounds from inorganic compounds that were likely present in the earliest of Earth's atmospheres. They put water, methane, ammonia, and hydrogen gases into a container and introduced an energetic spark. At the end of a week, 10 to 15 percent of the carbon within the system was now in the form of organic compounds and 2 percent of the carbon had formed amino acids. Whether or not the chemical soup they tested was representative of the Earth's early atmosphere, they had shown the ease with which organic compounds could be synthesized.

atmosphere would be expected to quickly escape, the Earth's more permanent early atmosphere was likely composed mostly of water vapor, nitrogen, and carbon dioxide. These gases probably arose as a result of outgassing from the planet's interior and from impacting near-Earth comets and asteroids. Earth could not have avoided the impacting objects that laid down a veneer of the organic materials necessary for life. When examined in laboratories, the meteorite fragments of some asteroids revealed plentiful organic compounds including amino acids, which are the building blocks of proteins, which are in turn the building blocks of living cells. A meteorite that fell near Murchison, Victoria, Australia, in 1969 was carefully examined and found to contain more than ninety different amino acids, nineteen of which are found in life on Earth. We, and all other forms of life, may owe our very existence to near-Earth objects.

Because there was no free oxygen in the Earth's atmosphere until about 2.4 billion years ago, early life forms developed a type of photosynthesis whereby they utilized the Sun's radiation to power their formation of digestible carbohydrates in reactions involving hydrogen, hydrogen sulfide, or iron. Other microbes produced methane, a greenhouse gas that helped keep Earth's surface waters liquid at a time when the Sun's energy output was well below current levels. About 2.7 billion years ago, some forms of algae evolved a more efficient type of photosynthesis that was capable of utilizing the Sun's energy to release oxygen into the atmosphere. Although oxygen was toxic to primitive microbes that had evolved in its absence, the continued production of oxygen reached a level whereby ozone could be formed. Ozone then helped block the Sun's ultraviolet radiation that was effective in breaking down atmospheric oxygen. This was an evolutionary leap forward for life on Earth because all nucleated cells require oxygen for metabolism.

The first multiple-cell organisms with DNA in their nuclei appeared more than 2 billion years ago. A few hundred well-preserved fossils discovered in Gabon during 2008 demonstrated that multicellular organisms existed at least 2.1 billion years ago. Much of the subsequent evolutionary action took place during a relatively short 50-million-year period that took place about 500 million years ago. It was during this so-called Cambrian explosion period when vertebrate animals, including the first fishes, appeared. Thereafter an immense variety of creatures on land emerged, punctuated by so-called mass extinction events that reset the evolutionary tracks for many species and allowed only the most adaptable to evolve further.

The Death of the Dinosaurs: Impacts Punctuate Evolution on Earth

For much of the history of paleontology, slow changes were thought to bring about the gradual evolution of different species. But more recently it was recognized that this slow evolutionary track was occasionally punctuated by so-called extinction events that took place in a geologically short period of time whereby one era ended and another era began with only the most adaptable species from the old era surviving into the new. Since K is the traditional abbreviation for the Cretaceous era or age of the reptiles and T represents the more modern Tertiary era, the extinction event sixty-five million years ago that marked this boundary has become known as the K-T extinction. While most groups of organisms survived, the K-T event saw the demise of most of the large vertebrates on land, in the sea, and in the air. Most plankton and many land plants were wiped out, but mammals, insects, and birds largely survived. At the time, mammals were relatively small creatures that fed upon larvae, worms, and snails, which in turn fed upon dead plants and animal matter. Amphibious crocodile-like reptiles and the majority of the sharks, rays, and skates also survived. But the large land reptiles that depended upon enormous food supplies of vegetation passed on, along with the carnivorous reptiles that preyed upon them. Very few fossils of these creatures are found above the K-T boundary. The dinosaurs were dead. What did them in?

In 1980 the father and son team of Luis and Walter Alvarez along with their colleagues published a paper in the journal *Science* claiming that a ten-kilometer-sized asteroid that collided with the Earth sixty-five million years ago was responsible for the K-T extinction event. An impact by a large asteroid would inject about sixty times the impactor's mass into the Earth's atmosphere with a fraction remaining in the stratosphere for several years. The resulting darkness would suppress photosynthesis, which would adversely affect plant life and the animals that depended upon these plants for food. The evidence for this astounding claim arose from their discovery of dramatic increases in the heavy metal iridium abundances in deep-sea limestone exposed in Italy, Denmark, and New Zealand at precisely the time of the K-T extinction event. Since most of the heavy platinum group metals, of which iridium is one, sank long ago toward the Earth's center, these metals are depleted in the Earth's crust relative to their cosmic abundances including the concentrations in the much smaller asteroids where the heavy elements, including iridium, would be expected to be distributed uniformly throughout their bodies. Within the one-centimeter sediment boundary layer between the Cretaceous and Tertiary sediments, the Alvarez team and their colleagues found iridium abundance increases of between 20 and 160 times the corresponding abundances in the older Cretaceous layer below this boundary layer and the younger Tertiary layer above it. The chemical composition of the boundary layer clay, composed of the settled stratospheric dust from the impactor, was markedly different from the clays mixed in with the Cretaceous and Tertiary limestones, which were chemically similar to each other. Today there are more than one hundred sites that have been found to exhibit this thin boundary layer with elevated iridium abundances.

A dramatic confirmation of the impact model for the K-T extinction event came with the discovery of the K-T extinction event crater buried beneath the Yucatán peninsula in Mexico. Evidence for an impact crater centered near the Mexican town of Chicxulub began to emerge in 1978 when the geophysicist Glen Penfield was helping to carry out a magnetic survey off the Yucatán peninsula in an effort to find suitable locations for oil drilling on behalf of the Mexican oil company Pemex. He found a huge feature that looked like a symmetric underwater arc. He also noted an arc-like feature in an earlier map that recorded the subtle differences in the region's gravity. The Canadian astronomer Alan Hildebrand had independent evidence for an impact in the region that included iridium-rich clays containing shocked quartz grains and small glass beads, all of which would be expected in

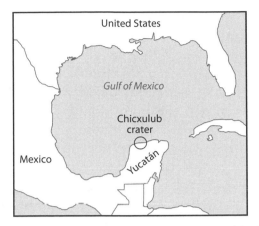

FIGURE 4.2. Map showing Chicxulub crater area on the edge of the Mexican Yucatán peninsula.

the ejecta resulting from an impact event. In 1990 Hildebrand contacted Penfield and the two then secured drill samples from the area taken by Pemex in 1951 and kept in storage ever since. The samples revealed shocked quartz that is indicative of an impact event of enormous pressure. The crater was dated to about sixty-five million years ago and its crater size of 180 kilometers or larger is consistent with the impact of a relatively large near-Earth object. Most scientists consider the Alvarez paper of 1980 and the Hildebrand paper of 1991 to be rather definitive evidence that the K-T extinction event was caused primarily by an impacting near-Earth object ten kilometers or larger.[7]

The K-T event was not the only extinction period to take place in Earth's history. For example, the transition from the Permian to the Triassic era about 250 million years ago is thought to have seen the extinction of more than 90 percent of marine species and 70 percent of land

[7] Agreement on the impact origin of the K-T extinction event is not universal. Some scientists consider a period of increased volcanic activity about sixty-five million years ago in west-central India to be a possible contributing event. The Deccan traps of India are some of the largest volcanic lava fields on Earth. The ramifications of enormous volcanic eruptions could include acid rain, ozone deletion, and climate change. It may be that a combination of an asteroid impact and increased volcanic activity were both responsible for the K-T extinction event. More recently, many researchers have begun referring to the K-T extinction event as the K-Pg extinction event (Cretaceous-Paleogene).

species. It is often referred to as the era of "The Great Dying," but if an impacting near-Earth object caused it, there is no obvious crater remaining from this event. However, it seems likely that an impactor would have hit an ocean and there are no ocean floor crusts older than two hundred million years because of sea-floor spreading and subduction.

The study of near-Earth objects is key to understanding the conditions under which the solar system formed 4.6 billion years ago. In terms of their chemical composition and thermal history, many of them are among the least changed members of our solar system. If we wish to understand the thermal environment and the chemical mix from which our solar system formed, then we need to study these small, primitive members of our solar system. They also warrant study for the clues they might provide concerning our origins. To what extent did the collisions of near-Earth objects provide Earth's water and the organic compounds that are the building blocks of life itself? Throughout geologic history, the Earth has been subjected to countless impacts by near-Earth objects, and some of the larger impacts created extinction events that punctuated the evolutionary process and allowed only the most adaptable species to evolve further. Sixty-five million years ago, mammals profited when the less adaptable large reptiles were rendered extinct by a ten-kilometer-sized asteroid hit. We humans likely owe our very existence and our preeminent position atop the world's food chain to collisions of near-Earth objects.

Until the mid-1990s, the known population of near-Earth objects was very modest. If these near-Earth objects are of such great importance, why were they not recognized until relatively recently? In the next chapter, we'll find out.

✦ **CHAPTER 5** ✦

Discovering and Tracking
Near-Earth Objects

No meteorite large enough to cause catastrophe would
ever again be allowed to breach the defences of Earth.
So began Project Spaceguard.
—Arthur C. Clarke, *Rendezvous with Rama*, 1973

Periodic active comets are considered near-Earth objects if they pass within 1.3 AU of the Sun and some of them, including comets Halley, Tempel-Tuttle, and Swift-Tuttle, have been recorded in ancient Chinese documents. Near-Earth comets have been known for a long time. For example, the 164 B.C. return of comet Halley was recorded on Babylonian clay tablets that are now housed in the British museum.[1] But comets are show-offs. When they enter into the inner solar system, their ices begin to vaporize and the resulting gases and the entrained dust particles stream away in an anti-Sunward direction, sometimes causing a visually spectacular display. Active comets may be show-offs but they make up only about 1 percent of the total near-Earth object population. While near-Earth asteroids don't belch gas and dust, it is these stealth asteroids that dominate the near-Earth object population and only recently have astronomers realized the immensity and importance of this population.

---

[1] See D. K. Yeomans, *Comets: A Chronological History of Observations, Science, Myth and Folklore* (New York: John Wiley and Sons, 1991), 265.

Filling the Gap between Mars and Jupiter

In the late sixteenth century, the German mathematician and astronomer Johannes Kepler suggested that the rather large gap between the orbits of Mars and Jupiter should contain an undiscovered planet. Kepler's suggestion was based largely upon his desire for symmetry and proportion in nature and the solar system. In 1766 the German astronomer Johann Daniel Titius suggested an interrupted empirical progression to identify the planetary distances from the Sun. If the distance from the Sun to the innermost planet Mercury is 0.4 AU, then the distance to Venus is 0.4 + 0.3 = 0.7 AU, Earth's distance is 0.4 + 0.6 = 1.0 AU, Mars's distance is then 0.4 + 1.2 = 1.6 AU, the next (missing) planet should then be at a distance of 0.4 + 2.4 = 2.8 AU, followed by the distances of the then outermost planets, Jupiter and Saturn, at 0.4 + 4.8 = 5.2 AU and 0.4 + 9.6 = 10.0 AU. The second component in each of these sums is doubled to arrive at the next planet's distance. Johann Bode, then director of the Berlin Observatory, used Titius's expression to argue the case for an undiscovered planet between the orbits of Mars and Jupiter and the expression became known as Bode's law. It was neither Bode's discovery nor a law, but when William Herschel discovered the planet Uranus in 1781 at a distance close to 19.2 AU, which was nearly equal to the Bode's law value of 19.6 AU, the Hungarian Baron Franz von Zach was convinced that Bode's law was correct.[2] According to von Zach, somewhere between the orbits of Mars and Jupiter there was a missing planet orbiting the Sun in the ecliptic region with all the other planets. After several years of fruitless searching himself, the Baron organized a systematic search of the heavens for the missing planet. In September 1800, he enlisted twenty-four astronomers who each agreed to search a specific section of the ecliptic region 15 degrees in length and 7–8 degrees in width. They were asked

---

[2] Bode's law is a nice confirmation of Stigler's law, "no scientific discovery is named after its original discoverer." See chapter 2, note 2. Bode's law does fairly well out to Uranus but not so well for Neptune. The actual semi-major axes, given first, as well as those predicted by Bode's law are: Mercury (0.4, 0.4), Venus (0.7, 0.7), Earth (1.0, 1.0), Mars (1.5, 1.6), Jupiter (5.2, 5.2), Saturn (9.5, 10.0), Uranus (19.2, 19.6), Neptune (30.1, 38.8). Bode's law is a nice example of a curious numerical relationship—but one without a sound physical basis.

to search for the missing planet and construct a star chart of their particular section. Letters outlining von Zach's plan were also circulated to other astronomers who were not involved with the celestial police effort. Surely such a well-organized effort by some of the best astronomers of the day would track down the missing planet.

One of von Zach's letters reached Father Giuseppe Piazzi, director of the Palermo Observatory in Sicily, who was already in the process of checking an existing star catalog. While not involved with the celestial police effort, Piazzi was systematically checking the position accuracies of each star in the catalog. On the evening of January 1, 1801, the first day of the nineteenth century, Piazzi noted that a star in the constellation Taurus he had observed the night before had moved! Because it lacked the diffuse appearance of a comet, Piazzi dared to hope that it might be something better than a comet. During several nights in January, Piazzi continued to follow its motion among the fixed stars. Piazzi followed the object until February 11. But as a result of poor weather, then an illness affecting Piazzi, and his reluctance to provide too much information before he had a chance to compute an orbit for his object, Bode did not receive news from Piazzi until March 20. The observations were not published and made available to the wider astronomical community until the summer of 1801. By this time, the new discovery had several alternative orbits and predicted positions were widely disparate. Piazzi's treasure, which he had named Ceres Ferdinandea (later shortened to Ceres), was effectively lost.[3] However, with characteristic genius, the young German mathematician Carl Gauss developed the technique necessary to determine an accurate orbit for Ceres and published his results in November 1801.[4] The object's semi-major axis, as computed by Gauss, was 2.77 AU, in apparent agreement with Bode's law. On December 7, 1801, von Zach observed Ceres at the German Seeburg Observatory, and subsequent observations through the end of the month confirmed that Ceres had

---

[3] Piazzi initially suggested the name Ceres Ferdinandea to honor King Ferdinand IV of Naples and Sicily.

[4] Although mathematically refined for modern computers, the initial orbit determination technique developed by Gauss to solve for Ceres's orbital elements in November 1801, subsequently published in 1809, is still in use today.

FIGURE 5.1. Portrait engraving of Giuseppe Piazzi (1746–1826), an Italian Catholic priest and astronomer who discovered the first asteroid, Ceres, on January 1, 1801, at the Palermo Observatory in Sicily.

been recovered. Although at the time Ceres was considered the missing planet between the orbits of Mars and Jupiter, what Piazzi had discovered was the first of the innumerable bodies we now call minor planets or asteroids, and the largest one at that.[5]

As so often happens in science, once the first of a new category of objects is discovered, many more discoveries quickly follow. The second asteroid, (2) Pallas, was discovered in March 1802, followed by (3) Juno and (4) Vesta in September 1804 and March 1807. The discovery of the fifth asteroid, (5) Astraea, would have to wait another thirty-eight

[5] Although the International Astronomical Union prefers the term "minor planets," it was William Herschel who first suggested the term "asteroids" because their appearance resembled small stars. Currently the terms "minor planets" and "asteroids" are used interchangeably. For the first half of the nineteenth century Ceres was considered a planet. Thereafter, it was termed a minor planet, and in 2006 the International Astronomical Union promoted Ceres from a minor planet to a dwarf planet.

years, but thereafter the discovery rate of asteroids in what came to be called the main belt between the orbits of Mars and Jupiter increased dramatically. Some astronomers who were trying to observe faint stars became irritated by an asteroid's often unexpected appearance in their star fields; one German astronomer derisively referred to them as the "vermin of the skies." Today more than three thousand main-belt asteroids are discovered each month.

Discovering the Near-Earth Asteroids: The Pioneers

But what about the discovery of near-Earth objects? Near-Earth objects include the Apollos, the Amors, the Atens, and the so-called inner Atiras, or inner-Earth objects. The first near-Earth asteroid to be discovered, in an Amor type orbit, was (433) Eros, discovered by both Gustav Witt at the Berlin Observatory and Auguste Charlois at the Nice Observatory on August 13, 1898. Witt took a two-hour exposure with a photographic plate all the while guiding the telescope at the same rate that the Earth rotated, hence keeping a certain set of stars immobile in his field of view. He noted that one object had a short trail on the photographic plate whereas the background stars were nearly point sources. When an orbit was computed, Eros's perihelion distance

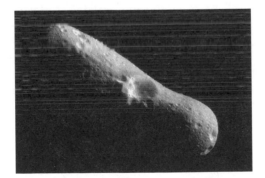

FIGURE 5.2. (433) Eros was the first near-Earth asteroid discovered. This image was taken by the NEAR Shoemaker spacecraft while it was in orbit around Eros in 2000. The longest dimension of Eros is 34 kilometers.
Source: Courtesy of NASA and the Applied Physics Laboratory of Johns Hopkins University.

was determined to be well inside the orbit of Mars. To honor this exceptional asteroid, the long-established custom of naming asteroids after women, both real and mythical, was abandoned—and some might argue the whole process of assigning names to asteroids has gone downhill ever since! Less than three years after its discovery, the Austrian astronomer Theodor Oppolzer observed that the brightness of Eros varied with time and suggested that an asymmetric body rotating with a period of a few hours might explain the observations. Many years later, after extensive ground-based and then spacecraft observations in 2000, the shape of Eros is known to be elongated and its rotation period is 5 hours and 16 minutes. Shortly after its discovery, astronomers took advantage of the proximity of Eros to determine through parallax its distance in kilometers and thereby improved the value for the astronomical unit, which is approximately the mean distance between the Sun and the Earth's orbit.

After the discovery of the first near-Earth asteroid Eros, a few more were discovered in the early twentieth century. The next one was the Amor type object (719) Albert in 1911, but it was not observed well enough to secure its orbit so it became lost for eighty-nine years until it was rediscovered in 2000 by Jeffrey Larsen at the Spacewatch Survey. Orbital investigations by Gareth Williams at the Minor Planet Center in Cambridge, Massachusetts, soon indicated that the discoveries in 1911 and 2000 were one and the same object. The third and fourth near-Earth asteroids to be discovered, both in Amor type orbits, were (887) Alinda in 1918 and (1036) Ganymed in 1924.

The year 1932, at the height of the Great Depression, was not a particularly good year—except for finding near-Earth objects. (1221) Amor was discovered by Eugène Delporte at Uccle, Belgium, on March 12, 1932, and (1862) Apollo was discovered on April 24 of the same year by Karl Reinmuth at the Heidelberg Observatory in Germany. These became the namesake asteroids for the Amor and Apollo groups of near-Earth asteroids. Then, in 1976, Eleanor Helin discovered the first near-Earth asteroid to have a semi-major axis less than that of the Earth. This object, (2062) Aten, then became the prototype for subsequent asteroids with this type of orbit. Atens and Atiras are among the most difficult to discover from ground-based observatories since they spend

Figure 5.3. Eleanor "Glo" Helin (1932–2009) was one of the early discoverers of near-Earth objects. She worked tirelessly to bring near-Earth objects to the attention of the astronomical community and the public.
Source: Courtesy of NASA/JPL-Caltech.

most or all of their time interior to the Earth's orbit and hence appear too close to the Sun to easily observe.

Eleanor ("Glo") Helin and Gene Shoemaker had begun an organized photographic campaign to discover near-Earth asteroids in January 1973 using the eighteen-inch aperture Schmidt telescope at Palomar Mountain in Southern California.[6] Glo did most of the observing and managed to take a pair of twenty-minute and ten-minute exposures for each of seven regions of the sky per night. This program

[6] Eleanor Helin was a pioneer in the hunt for near-Earth objects and was the first woman observer allowed to regularly use the on-site Palomar Observatory lodging facility, which was dubbed the "Monastery." Until Glo crashed the gender barrier, the Monastery was an all boys' club. Glo, one of the most determined people I have ever met, simply would not tolerate such discriminatory behavior—tradition be damned. She had a heart of gold but she could be feisty. In 1982, after a bitter argument with Gene Shoemaker during an observing session at Palomar, she returned to the Monastery and locked the outside door before Gene had a chance to get in and go to sleep. Their cooperative arrangement ended shortly thereafter. See David Levy, *Shoemaker by Levy: The Man Who Made an Impact* (Princeton: Princeton University Press, 2000), 172.

FIGURE 5.4. Gene and Carolyn Shoemaker worked together to discover many near-Earth asteroids and comets at the Palomar Observatory.
Source: Courtesy of Glen Marullo.

was called the Palomar Planet-Crossing Asteroid Survey (PCAS). Compensating for the Earth's rotation, the stars were tracked across the sky and each pair of photographic films was then checked by microscopes for the telltale image streaks of near-Earth asteroids among the point-like star images. Success was slow. The first discovery was not until six months later when the highly inclined Apollo asteroid (5496) 1973 NA was discovered on July 4, 1973. By 1978 they had discovered twelve near-Earth asteroids, thus doubling, in five years, the total discoveries made since Eros was first found in 1898. By 1980 Shoemaker and Helin had abandoned the examination of image streaks in favor of a more efficient discovery technique. Using fast emulsion films, they began taking two short exposures made several minutes apart so images would be in slightly different locations on the two photographic films. The two images were then examined with a stereomicroscope. The near-Earth asteroid images, now slightly displaced points of light rather than streaks, seemed to float above the star background. Gene's wife, Carolyn Shoemaker, who was particularly talented at discovering near-Earth objects using the stereomicroscope, carried out much of this work. By 1980 some friction between Helin and the Shoemakers resulted in the dissolution of their cooperative program. Mom and Pop Shoemaker, as the Shoemakers jokingly referred to themselves, then relocated back to the U.S. Geological Survey in Flagstaff, Arizona, and Helin moved her separate photographic program to JPL. The Shoemakers continued their photographic search program, the Palomar Asteroid and Comet Survey (PACS), until the mid-1990s.[7]

CCD Detectors Revolutionize the Discovery of Near-Earth Objects

In mid-1983 Tom Gehrels and Bob McMillan began using the 0.9-meter aperture Steward Observatory telescope for near-Earth object search efforts. Located near Tucson, Arizona, this effort would become known as the Spacewatch Survey. Gehrels held the telescope fixed in its mount so the celestial objects would constantly scan through the telescope's field of view as a result of the Earth's rotation. By 1984 this telescope was being used full-time for near-Earth object searches and it employed, for the first time, a charge-coupled device (CCD). These detectors, now so common in astronomical applications, digital cameras, and cell phones, were new technology at that time for near-Earth object discovery efforts. Initially, the Spacewatch CCD had an array of 320 × 512 individual sensors (pixels), each of which built up an electric charge proportional to the light intensity on the focal plane at that location. By 1989 Spacewatch began full-time near-Earth object search operations using a CCD camera that was two thousand individual sensors (pixels) wide and two thousand pixels long (2k × 2k). Today the Spacewatch CCD camera detector system is larger still. The discovery technique involves taking images of the same star field several times a

---

[7] In July 1997 Gene Shoemaker was killed in a tragic automobile accident in Australia. He had been trained as a geologist, and one of his biggest disappointments was being disqualified from the Apollo astronaut program because be had Addison's disease, a rare condition affecting the adrenal glands that could be controlled with steroids. However, he was instrumental in teaching the Apollo astronauts about the lunar surface, and a portion of his cremated ashes were placed onboard the Lunar Prospector mission to the Moon. In July 1999, two years after his death, Gene did make it to the moon when the Lunar Prospector spacecraft, having completed its scientific mission, was commanded to impact the moon's surface.

Figure 5.5. Cartoon, from The Other Coast.
Source: Adrian Raeside.

night with each separate image taken fifteen to twenty minutes apart. A computer program then can compare the images and quickly identify and ignore the stars that don't move between images. Gehrels wryly referred to these stars, rather than the moving asteroids, as "the vermin of the skies." Objects that do move from image to image are bodies within our solar system and can then be separated into the most common slower-moving asteroids in the main belt and the rarer, rapidly moving near-Earth asteroids with their longer trails and differing motions on the star background.

All modern surveys for near-Earth objects use large-format CCD detectors rather than old photographic emulsions because of their ease of use, digital output, high light-gathering efficiency, and linear response to light intensity (e.g., light intensity ten times stronger in one pixel compared to a neighboring pixel will record ten times the electric charge).[8]

NASA Gets Serious about Finding Near-Earth Objects

The introduction of the CCD technologies allowed far more efficient discovery surveys to be carried out. The impetus for actually carrying out these surveys received a boost due to a workshop run by Gene Shoemaker in July 1981 in Snowmass, Colorado. Although never formally published, the circulated report from this workshop called attention to the unusual nature of asteroid impacts; although large Earth-impactor events are extremely rare, they are of very high consequence. The announcement a year earlier by the father and son Alvarez team that the death of the dinosaurs was likely due to an asteroid impact underscored this conclusion. The Snowmass report recommended that telescopic searches be carried out and stressed the importance of finding Earth-threatening objects early enough to initiate mitigation campaigns. In a particularly prescient fashion, the report also recommended that accurate collision prediction techniques be developed, deflection techniques be studied, and the physical nature of near-Earth objects be

[8] There are a number of observing programs in the international arena that have made significant contributions to the discovery of near-Earth objects and the critically important follow-up observations that are required to secure their orbits. These programs include a (discontinued) German and French effort using the 0.9-meter aperture telescope located at the Cote d'Azur Observatory north of Cannes, France, an Italian-German collaboration located at Asiago-Cima Ekar, Italy, the Klet Observatory in the Czech Republic, and the Japanese Spaceguard Association's program at Bisei, Japan. Mention should also be made of Alan Fitzsimmons, who uses the one-meter telescope at La Palma in the Canary Islands, and Peter Birtwhistle, who is a remarkably successful follow-up observer despite working in England—a land not noted for its clear skies. In the United States, some of the leading follow-up observers include Dave Tholen in Hawaii, Bob McMillan, Jeff Larsen, Jim Scotti, and Terrence Bressi at the Spacewatch Observatory near Tucson, Bill and Eileen Ryan, who use the 2.4-meter aperture telescope at the Magdalena Ridge Observatory in New Mexico, Alice Monet and Hugh Harris at the Naval Observatory near Flagstaff, Arizona, a group led by Robert Holmes in Illinois, and Bill Owen and Jim Young (now retired) at JPL's Table Mountain Observatory in Southern California.

investigated to "gather the engineering data base now lacking for appropriate design of collision avoidance mechanisms."

Recommendations of this type in the 1980s and early 1990s did not receive immediate or wide acceptance and there was, for a time, a certain "giggle factor" that the media and some members of the scientific community associated with the threat of Earth-colliding asteroids. It was difficult to make a case that a disaster that no one had ever experienced was worth worrying about. In an influential paper published in 1994, Clark Chapman and David Morrison outlined the nature of the threat to Earth from near-Earth asteroids and comets. They pointed out that although large-impact events are beyond our personal experience, these events when they do occur are so catastrophic that their long-term statistical hazard is comparable to that of many other, more familiar disasters like aircraft crashes, floods, and tornadoes. The lower limit for the diameter of an impacting asteroid that could cause a global rather than a regional disaster was identified as about 1.5 kilometers.

In 1990 the U.S. House of Representatives in their NASA Multi-year Authorization Act directed NASA to undertake two workshop studies to evaluate the threat to the Earth of asteroid and comet impacts and to explore remedial actions that would prevent such disasters. The first of these two reports was issued in early 1992; the NASA Spaceguard Survey Report, chaired by David Morrison, included the goal of discovering more than 90 percent of near-Earth objects (NEOs) larger than one kilometer within twenty-five years. Two years later, the U.S. Congress House Committee on Science and Technology passed an amendment to the NASA authorization bill directing NASA to report within a year with a program to identify and catalog, within ten years, all NEOs larger than one kilometer.[9] In 1995 another NASA panel of experts, chaired by Gene Shoemaker, recommended two dedi-

[9] The second of these two reports, titled "Summary Report of the Near-Earth-Object Interception Workshop," was issued in August 1992. NASA's John Rather and Jurgen Rahe chaired this workshop. The workshop report noted that there existed technically credible approaches to preventing most impact catastrophes provided that the requisite technology capabilities were developed and an appropriate experimental program is eventually undertaken. Some participants urged that tests be conducted in space to alter the course of small asteroids by using impacts by spacecraft that did not carry explosives. However, a number of participants expressed great concern regarding the use of nuclear explosions in space.

cated telescopes of two-meter aperture and one or two one-meter telescopes with advanced focal plane detectors to accomplish this goal.

For several years there was some interest in near-Earth objects expressed by members of Congress, particularly George E. Brown, congressman from California and chair of the House Science Committee. The effort to discover near-Earth objects received a boost in May 1998 when Carl Pilcher, then the science director for the Solar System Exploration Office of Space Science, announced at a congressional hearing that NASA would undertake a survey program with the goal of discovering 90 percent of the near-Earth objects larger than one kilometer within ten years. This would become known as NASA's Spaceguard Goal.[10] In summer 1998, NASA initiated the Near-Earth Object Observations Program to detect, track, and physically characterize this population. Tom Morgan of NASA's Solar System Exploration Division managed this program for several years. This goal was driven by the conclusion that one- or two-kilometer-sized objects are at the lower limit where an impact could have global rather than just regional consequences. While impacts of this large an object are now very rare, say once every seven hundred thousand years on average, the consequences of such an event would be so great that they, rather than the far more frequent impacts by smaller objects, would represent the greatest long-term threat to humans.

The Spaceguard Goal is expressed in terms of NEOs of a certain size. However, astronomers using optical telescopes cannot directly determine the sizes of asteroids, so an assumption as to their reflectivity, or albedo, must be made to estimate a size. If we assume that NEOs are spheres reflecting only about 14 percent of the sunlight incident upon them, then their sizes can be inferred since their apparent brightness is observed and their distances from both the Earth and Sun can be calculated from orbital computations.

The Lincoln Near-Earth Asteroid Research (LINEAR) program run by MIT's Lincoln Laboratories has discovered the most NEOs larger

---

[10] The term "Spaceguard" originated in Arthur C. Clarke's 1973 science fiction novel *Rendezvous with Rama*. In this novel, Project Spaceguard was established to identify near-Earth objects on Earth-threatening trajectories.

than one kilometer. Operating near Socorro, New Mexico, the LIN-EAR survey observes with two co-located one-meter aperture tele-scopes that were formerly operated by the air force for Earth-orbiting satellite surveillance but are now upgraded to use CCD imaging devices that are capable of being read out very quickly, thus increasing their survey efficiency.

The current frontrunner in terms of annual near-Earth object discoveries is the Catalina Sky Survey near Tucson, which operates a 0.74-meter telescope on Mt. Bigelow, 1.0- and 1.5-meter aperture telescopes on nearby Mt. Lemmon, and a half-meter aperture telescope near Siding Spring, Australia, that was in operation from 2004 through 2011.[11]

If the Spaceguard Goal is to find 90 percent of the population of one-kilometer and larger near-Earth objects, how do we know how many of these large objects are in the total population? Another approximation is in order. Imagine that you, along with your astronomer colleagues, are observing near-Earth objects with telescopes over several years and your celestial police force succeeds in discovering 100 near-Earth objects. Hooray for you—but wait. It then turns out that 90, or 90 percent, of these "discoveries" had already been reported. They are actually re-discoveries. You could then assume that 90 percent of the entire population had already been discovered and if there were already about 900 actual near-Earth object discoveries, you could then assume that the total population (100 percent) would be 1,000 objects larger than one

[11] Grant Stokes is the initial Principal Investigator for the LINEAR project with co-investigators Scott Stuart, Eric Pearce, and Michael Harvanek. Started by Steve Larson in 1998, the Catalina Sky Survey has been run by Ed Beshore as the Principal Investigator. The highly successful observing team includes Andrea Boattini, Gordon Garradd, Alex Gibbs, Al Grauer, Rik Hill, Richard Kowalski, and Rob McNaught. Other discovery surveys, now discontinued, include a short-lived photographic search program led by Duncan Steel in 1990 using the 1.2-meter Schmidt telescope at Siding Spring, Australia. Ted Bowell and Larry Wasserman ran the Lowell Observatory NEO Survey (LONEOS) from 1993 through 2008 using the Lowell Observatory 0.6-meter telescope. Beginning in 1995 and continuing for several years, the JPL Near-Earth Asteroid Tracking (NEAT) program, in cooperation with the air force, ran a one-meter then a 1.2-meter air force telescope on Hawaii's Mt. Haleakala for survey purposes. In 2001 the NEAT program began a transition to using the 1.2-meter Schmidt telescope at Palomar Mountain in Southern California. Begun under the direction of Eleanor Helin, the Principal Investigator, and David Rabinowitz, the NEAT program was led successively by Steve Pravdo and Ray Bambery at JPL and then by Caltech's Mike Brown before ceasing operations in 2007. JPL's Ken Lawrence was active in the data reduction process throughout the twelve-year NEAT survey as well as during some of the earlier photographic survey work carried out by Glo Helin.

kilometer in diameter. The actual number of near-Earth asteroids larger than one kilometer is about 990 give or take a few dozen either way, according to the astronomer Alan W. Harris (the elder), who carried out a far more sophisticated analysis, but this is the general idea.[12]

Interactions of the Minor Planet Center and the JPL and Pisa Trajectory Computation Centers

All of the positions relative to the stars, from which orbits can be calculated for NEOs, as well as all other solar system bodies, are sent first to the Minor Planet Center (MPC) in Cambridge, Massachusetts. The MPC is an organization sanctioned by the International Astronomical Union, funded largely by NASA, and currently under the direction of Tim Spahr. The MPC collects these data, verifies them, provides object designations and discovery credits, and makes the data available to the public, including the trajectory computation center at the Jet Propulsion Laboratory (JPL) in La Cañada, California, and at an Italian and Spanish Near-Earth Object Dynamic Site (NEODyS) run out of the University of Pisa, Italy, and the University of Valladolid, Spain. The MPC computes preliminary orbits and then looks to see if additional observations of this object are available in their archives that can be used to refine the preliminary orbit. The MPC has many additional responsibilities including the generation of preliminary orbits for NEOs, web-based notification of potential new NEO discoveries to follow-up observers, and the generation of ephemerides, or tables of future position predictions, that enable these follow-up observations. They successfully process enormous quantities of observations from the

---

[12] Alan W. Harris carried out a so-called size-frequency study in 2006 and 2011 to determine how many near-Earth asteroids (NEAs) there are at a particular size. For example, he determined that there are about 990 NEAs larger than one kilometer, about 20,000 NEAs larger than 140 meters, and more than a million NEAs larger than 30 meters in diameter. As the large asteroids in the main belt bang into each other over millions and millions of years, smaller and smaller fragments result. One then winds up with lots of small ones and relatively few large ones. The same thing happens when you hit a brick with a sledgehammer—lots of little pieces and a few larger ones. Believe it or not, there are two asteroid scientists named Alan William Harris, one in Southern California and the other in Germany. The Alan Harris referred to here lives in Southern California and is seven years older than his German colleague, so we refer to him as Alan W. Harris, the elder.

international observing community and announce new discoveries. More than three thousand main-belt asteroids and about eighty near-Earth objects are discovered each month.[13] Once the MPC has gathered the observational data and done preliminary orbits for near-Earth objects, they electronically transmit these data to the trajectory computation centers located at the JPL and in Pisa.

At JPL, once the data are received, an automatic orbit determination and prediction process is conducted; information on future close-Earth approaches is made immediately available on the JPL NEO website. If a particularly close approach to Earth within the next century or so is noted as being possible by the software system, the object enters the Sentry system, which computes potential future Earth-impact probabilities and associated information such as impact time, relative velocity, impact energy, and impact hazard scale values. Sentry alerts are automatically posted to the NEO Program Office website (neo.jpl.nasa.gov). For objects with relatively high impact probabilities, high impact energies, or short intervals until the time of impact, the Sentry system will automatically notify the office staff for manual verification before posting the results on the web. In these latter cases, the results are first checked for accuracy and then sent to the NEODyS personnel for verification. At the NEODyS facilities, a similar process will have already been under way, and if both the JPL Sentry system and the NEODyS system yield equivalent results, the relevant information is posted on both the JPL and NEODyS websites nearly simultaneously. Since the Sentry and NEODyS systems are independent, this cross-checking provides a valuable verification process before the publication of information on high-interest objects.[14]

[13] Tim Spahr, Gareth Williams, Jose Galache, Sonia Keys, and Carl Hergenrother work tirelessly to process these data. Prior to his untimely death in November 2010, Brian Marsden, director of the MPC from 1978 to 2006, also did orbital computations for the comets including the small suicidal Solar and Heliospheric Observatory (SOHO) comets that keep colliding with the sun. Launched in December 1995, the SOHO spacecraft was designed to monitor the sun's activity. SOHO images have also been used to discover more than two thousand small comets near the sun. Many of these discoveries have been made by an international group of amateurs searching through the publically available SOHO images online: http://sungrazer.nrl.navy.mil/index .php?p=cometform.

[14] At Pisa, Andrea Milani is in charge; he is aided by Giovanni Gronchi, Fabrizio Bernardi, Giovanni Valsecchi, and Genny Sansaturio at the University of Valladolid. Don Yeomans is the manager of NASA's NEO Program Office at JPL; key personnel include Steve Chesley, Alan

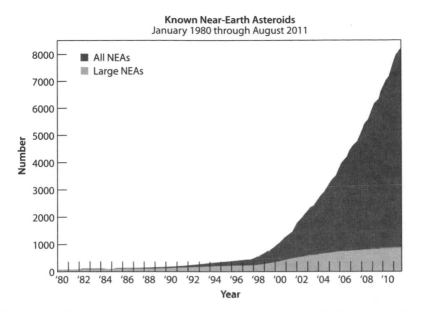

FIGURE 5.6. Plot showing the rapid increase in near-Earth asteroid discoveries in the late 1990s and beyond. The darker curve shows the number of discoveries of near-Earth asteroids of all sizes versus date while the lighter curve shows the number of discoveries of one-kilometer and larger near-Earth asteroids versus date.
Source: Courtesy of Alan Chamberlin, NASA/JPL-Caltech.

Lindley Johnson at NASA headquarters, Planetary Science Division, in Washington, D.C., now runs the entire Near-Earth Observations Program, including the discovery surveys, the follow-up observations, and the physical characterization of these objects.

The Next Generation of Near-Earth Object Search Surveys

In 2003 NASA released a Near-Earth Object Science Definition Team report recommending that near-Earth object searches be extended to

---

Chamberlin, Paul Chodas, and Jon Giorgini. Steve Chesley was a key contributor in setting up both the NEODyS system in Italy, which came online in January 1999, and the Sentry system at the JPL, which came online two years later.

smaller objects whose diameters are 140 meters and larger.[15] This new target size goal was selected because it represents the level of survey needed to reduce by 90 percent the risk from unwarned sub-kilometer impactors. With the Spaceguard Goal having already reduced the global undetected risk by more than a factor of ten, completion of the new survey to the 90 percent level would reduce the total assessed risk from undetected objects of all sizes to less than 1 percent of the risk prior to the 1998 start of the NASA survey efforts. Impacting asteroids larger than 140 meters in diameter would be expected to punch through the Earth's atmosphere and either cause significant regional destruction with a land impact or create damaging tsunamis with a water impact. Successful surveys reduce the hazard from undiscovered near-Earth objects in the sense that knowledge of an impending impact would allow sufficient time to use existing technologies to mitigate the risk.

Examples of next-generation ground-based search surveys include the Panoramic Survey Telescope and Rapid Response System (PanSTARRS) and the Large Synoptic Survey Telescope (LSST). The PanSTARRS is being built with development funding from the U.S. Department of Defense. The PanSTARRS 1 telescope, which began operations in 2010, is a single 1.8-meter aperture telescope operating on the Haleakala peak in Maui. The plan is to take CCD images of patches of sky (7 square degrees) twice each evening and cover the entire accessible sky three times per lunar month (28 days) using their newly developed, very large format, 1.4 giga-pixel CCD camera.[16] A moving object will receive two observations during the first discovery evening and a set of two additional observations for another two nights within each 28-day period. Future plans call for a second 1.8-meter telescope and possibly a PanSTARRS 4 telescope with four co-located 1.8-meter aperture telescopes. The latter system should be able to image sky fields with four times the sensitivity of the single-telescope PanSTARRS 1 system, which is planned to routinely survey to visual magnitude 22.[17]

---

[15] See the 2003 report at: http://neo.jpl.nasa.gov/neo/report.html.

[16] For comparison, the angular area of the full moon is about 0.2 square degrees.

[17] Astronomical magnitudes are used to measure the apparent brightness of a celestial body. The brightest star, other than our sun, is Sirius, which has an apparent magnitude of –1.5. Each

The U.S. National Science Foundation, the U.S. Department of Energy, private donors, and a number of additional academic and institutional sponsors are funding the LSST. The planned aperture is 8.4 meters in diameter with a field of view of 9.6 square degrees. It will be located at Cerro Pachon, in northern Chile, and if the requisite additional funding can be secured, it is planned for "first light" in 2018. The observing plans are to cover the entire accessible sky every three nights down to fainter than apparent magnitude 24.7.[18]

While neither PanSTARRS 1, PanSTARRS 4, nor LSST will be devoted solely to the study of near-Earth objects, all of these programs have included NEO discovery as a primary science goal. The product of a search telescope's field of view multiplied by the aperture area of the telescope is often used as a metric for the efficiency with which a survey can discover NEOs. This product is approximately 2 for the best-performing discovery system currently in operation. If these proposed systems achieve their goals, the corresponding values for PanSTARRS 1, PanSTARRS 4, and LSST will be approximately 12, 51, and 319, respectively. The LSST system operating in its designed observing mode would be expected to find 90 percent of the near-Earth objects larger than 140 meters in about seventeen years of operations. A LSST system dedicated to finding near-Earth objects could do the job within twelve years of starting operations.

In late December 2005, NASA was directed by Congress to recommend an option to carry out the survey program for objects 140 meters and larger. A year later, NASA responded that the agency could reach the new goal by partnering with other agencies on potential future optical ground-based optical observatories and building a dedicated survey system.[19] Space-based infrared telescopes would be a more efficient discovery option because the radiation from dark asteroids peaks

---

increase of a single magnitude represents a drop in brightness by a factor of 2.5 so a magnitude 6 star would be 100 times fainter than a magnitude 1 star.

[18] The progress of the LSST can be followed at: http://www.lsst.org/lsst_home.shtml.

[19] See NASA's 2007 report at: http://neo.jpl.nasa.gov/neo/report2007.html. The prestigious National Research Council endorsed many of the 2007 report recommendations in a subsequent report in 2010 titled *Defending Planet Earth: Near-Earth-Object Surveys and Hazard Mitigation Strategies* (Washington, DC: National Academies Press, 2010).

in the infrared and the view from an observatory orbiting the Sun interior to the Earth's orbit would have far better viewing coverage of the most dangerous objects that would be expected to be in Earth-like orbits. In addition, a space-based telescope would not have to deal with the downtime due to weather and daylight.

Simulations by Steve Chesley at JPL and by Roger Linfield at the Ball Aerospace Corporation have indicated that a half-meter, wide-field telescope operating at near-infrared wavelengths in a Venus-like orbit would be capable of reaching the goal of finding 90 percent of the near-Earth objects larger than 140 meters in just over eight years when operating by itself, in about six years if operated in conjunction with the PanSTARRS 1 system, or about three years if operated in conjunction with an LSST system dedicated to the search. A space-based infrared telescope in an orbit interior to that of the Earth would be the most efficient discovery system for near-Earth objects. On the other hand, ground-based optical surveys would be less expensive and last longer because of the ease with which they can be maintained. The most efficient and robust discovery surveys would include a combination of space-based and ground-based assets.

The Wide-field Infrared Survey Explorer (WISE) spacecraft was launched on December 14, 2009, and operated for ten months in 2010 before its coolant for the infrared camera was exhausted. It then operated for four more months of "warm" operations to complete two sweeps of the celestial sky. Although the telescope was not designed for the discovery of near-Earth objects, co-investigator Amy Mainzer and her team successfully operated a program to discover near-Earth objects and comets by mining all the infrared images collected by WISE to detect objects that moved from frame to frame. In fourteen months of operation, they managed to get discover credit for 135 previously unknown near-Earth objects and 21 comets. This so-called NEOWISE program demonstrated the facility with which space-based infrared telescopes can discover near-Earth objects. NEOWISE would have been credited for discovering several more near-Earth objects, but some of the potential discoveries were lost because suitably large ground-based optical telescopes were not available to provide the follow-up observations necessary to secure their orbits.

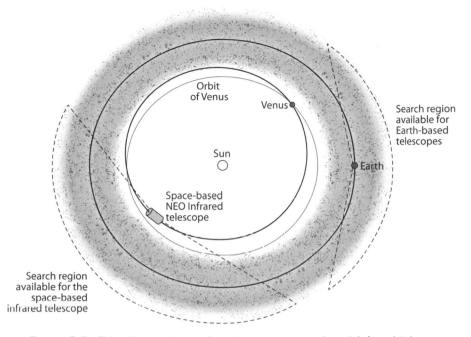

FIGURE 5.7. This diagram shows the advantage a space-based infrared telescope would have over Earth-based optical telescopes for discovering near-Earth asteroids. The nominal search region available from Earth is shown, where Earth is located at the 3 o'clock position. This search region should be compared to the much larger region available from an NEO infrared telescope located at the 8 o'clock position on an orbit similar to that of the planet Venus, interior to the Earth's orbit about the Sun. In addition, when compared to optical telescopes on Earth, this space based infrared telescope would more easily see dark near-Earth asteroids, would circulate around the Sun more rapidly than the Earth, could operate twenty-four hours per day, and would have less confusion from the background stars.

Since the NEOWISE observations are made in the infrared region of the spectrum, they are measuring the heat radiating from an asteroid rather than its reflected sunlight or brightness, which is measured by telescopes operating in the visible region of the spectrum. As a result, NEOWISE observations are able to infer an asteroid's size to an uncertainty of only about 10 percent and its reflectivity to about 20 percent. Whereas optical observations of most asteroids must assume a reflectivity to determine an approximate size, the NEOWISE observations allow a much more accurate determination of size and reflectivity. The NEOWISE observations suggest that the numbers of near-Earth objects with sizes in the range of one hundred to a few hundred meters are systematically lower than the totals inferred using only the optical observations. Thus using the NEOWISE observations, the total number of near-Earth asteroids with diameters larger than one kilometer, 500 meters, 140 meters, and 100 meters are about 980, 2,400, 13,000, and 20,500, respectively. Using the optical observations and assuming an average reflectivity of 14 percent, the corresponding totals are about 990, 3,300, 20,000, and 36,000 (see table 8.1).

If a recently discovered near-Earth object is close enough to Earth to allow follow-up radar observations from either the Goldstone seventy-meter antenna in Southern California or the thousand-foot aperture antenna in Arecibo, Puerto Rico, then the object's orbit can be immediately secured. Radar observations in the form of line-of-sight distances (range) and line-of-sight radial velocities (Doppler) are extremely powerful data types when used in combination with optical observations. That is, while optical observations can define the object's position in a plane perpendicular to the observer's line of sight, radar observations can provide the third dimension in the line-of-sight direction and do so with the surprising accuracy of a few meters in range and a millimeter per second in radial velocity. According to a study by JPL's Jon Giorgini, an initial orbit that includes radar data will allow, on average, an accurate extrapolation of the object's motion five times further into the future than would be possible with an orbit based solely upon optical data, thus preventing the object from being lost or requiring rediscovery. However, once there are optical data available for two or more returns of the object to perihelion, the orbit improvement provided by the radar observations is not nearly so dramatic.

Radar observations can dramatically improve an object's orbit and, at the same time, provide enough information to determine its rotation characteristics and surface properties along with a shape resolution far greater than optical telescopes. In the next chapter, we'll take a look at what is known about the near-Earth comets and asteroids, via ground-based observations, spacecraft observations, and radar studies.

CHAPTER 6

The Nature of Asteroids and Comets

Asteroids are older than dirt.

Donald Duck and Uncle Scrooge McDuck Were There First

In 1960, before video games, cell phones, and the Internet, much of a youngster's entertainment came from comic books. Disney's Donald Duck and his fabulously wealthy Uncle Scrooge McDuck were comic book adventure heroes. In a 1960 comic book release titled "Islands in the sky," Donald Duck, Uncle Scrooge, and Donald's nephews Huey, Dewey, and Louie embarked upon a space adventure to the asteroid belt in a secondhand "skyfish spacewagon" rocket ship recently purchased by Uncle Scrooge. Their mission was to locate a bare, rocky little planet where Scrooge could store his money safely. During their adventures in the asteroid belt, they discovered some asteroids were no larger than houses and some had the oddest shapes. An asteroid with a moon was located and Donald Duck, on an extravehicular activity, dramatically jumped onto and through an asteroid composed of cinders and rocks that "wasn't even glued together!"

So there you have it. As far back as 1960, Donald Duck and his fellow astronauts determined that asteroids have widely diverse shapes; some have moons and some are porous rubble pile collections of material that have very little cohesiveness.[1] It would take a few decades before human planetary scientists would reach the same correct conclusions.

Rubble Pile Asteroids: Collisions Rule

In the early solar system, slow collisions of rocky particles allowed asteroids to accumulate material and grow into larger bodies. If these planetesimals were large enough, their own self-gravity would pull them into roughly spherical shapes. There is also overwhelming evidence that some asteroids were formed into layered structures with nickel-irons cores, surrounding silicate rock mantles, and rubble-like rocky surfaces formed by the ejected material from the cratering explosions of much smaller impacting asteroids. This evolution into a layered structure is called differentiation. Our Earth has been differentiated into a layered structure, and there is evidence that the large asteroid Vesta is also differentiated. Apparently these layered, or differentiated, asteroids were strongly heated in their youth and this heating melted the heavy metals, which then sank into the central core regions of the asteroid. It is not clear how this heat was generated in the early solar system, but one strong possibility is that it was generated by the radioactive decay of an isotope of aluminum (^{26}Al) into an isotope of magnesium (^{26}Mg). The unstable isotope ^{26}Al spontaneously decays to ^{26}Mg, emitting heating energy in the process. Most aluminum in asteroids is in the form of the stable ^{27}Al, but if only a tiny fraction of the aluminum were ^{26}Al, the decay process would generate enough heat to completely melt asteroids several kilometers across.

Figure 6.1 is a cartoon showing a catastrophic collision between two large asteroids, one differentiated and one solid rock. The collision forms smaller fragment asteroids that have diverse structures. They

[1] Carl Barks, a writer and illustrator, created this comic book story and many other Donald Duck comic books that featured space exploration and invention. In early 1983, Carl Barks was honored by having an asteroid named after him, (2730) Barks. Ted Bowell at the Lowell Observatory near Flagstaff, Arizona, discovered this asteroid and I am thankful to Ted for drawing my attention to this anecdote.

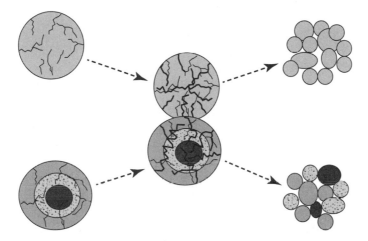

FIGURE 6.1. Cartoon showing the origin of different asteroid types and the meteorites that can arrive on Earth from these asteroids. A differentiated, or layered, asteroid could collide with an undifferentiated, or unlayered, rocky silicate body, forming smaller asteroids and potential meteorites. The collision fragments can have the characteristics of the original rocky silicate object or of the various regions of the differentiated asteroid, including the loose top layer, the stony-iron nature of the middle region, or the iron-nickel metallic nature of the core region.

include a host of small rocky bodies, stony-iron bodies, rubble piles, and a few solid nickel-iron chunks. One such metallic chunk, about fifty meters in diameter, slammed into the Arizona desert fifty thousand years ago, created Meteor Crater, and spread nickel-iron meteorites all around the crater's rim (see figure 4.1).

The strong radar reflections from some asteroids like the dumbbell-shaped (216) Kleopatra suggest a metallic surface, and the majority of the meteorites in collections today are metallic nickel-irons. This strongly suggests that their parent body asteroids had metallic cores before being broken apart during a long-ago catastrophic collision with another asteroid.[2] There is also ample evidence that some small

[2] Fragments of nickel-iron asteroids are tough and often survive passage through the Earth's atmosphere, so even though they are relatively rare in terms of numbers, nickel-irons are the most numerous objects in meteorite collections. They are also more easily recognized than the more numerous rocky meteorites. But nickel-iron asteroids are also rare in the near-Earth object

asteroids are solid rock and rocks fractured by collision. The large near-Earth object (433) Eros seems to be a fractured rocky body thirty-four kilometers in length. But most of the larger asteroids are likely rubble piles or objects held together by little more than their own self-gravity. So just what is a rubble pile asteroid?

Until the late 1970s the consensus opinion among planetary scientists was that asteroids were whirling rocky bodies. In 1977 the planetary scientists Clark Chapman and Don Davis demonstrated that some asteroids are likely to be loose collections of collision fragments, or rubble piles. They reasoned that over long time intervals, asteroids run into each other and the energy needed to break up a rocky asteroid during a collision is a lot less than the energy to completely disperse the resulting fragments. Thus if an asteroid were smashed to smithereens by a collision with another asteroid, it would be more likely to reassemble into a loose collection of fragments than to have all the fragments completely disperse into space.

Some scientists describe a rubble pile body as the type of structure that one would get if you dumped a truckload of rocks in a field. Others note that even if the individual rocky clumps were loose, they would likely have some tensile strength arising from the friction between the larger chunks and the electrostatic forces between the smaller particles. Dan Scheeres, a dynamicist at the University of Colorado, has pointed out that the very low asteroid gravity means the weight of surface particles is also very low. Hence until rubble pile chunks of rock exceed about one meter in diameter, the electrostatic forces between them (so-called Van der Waals forces) are equivalent to the gravitational attraction between them. As a result, sometimes even so-called rubble pile asteroids have a bit of strength. Van der Waals forces are also responsible for the strange structures that can form in bread flour without collapsing. Dan Scheeres notes that it may be that the gravels and rocks on the surfaces of small asteroids behave similarly to bread flour.

---

population. They are most often associated with M-type asteroids where "M" is often used as a stand-in for "metallic." However, not all M-type asteroids are metallic and some even have evidence for water in the form of hydrated minerals. These hydrated minerals can form when water molecules incorporate directly into the crystalline structure of a mineral, like feldspar, to form clay minerals.

FIGURE 6.2. Asteroid (253) Mathilde was imaged by the NEAR Shoemaker spacecraft in June 1997. Four huge craters, each with diameters comparable to Mathilde's radius, were identified in the spacecraft images. Mathilde's longest dimension is 66 kilometers. It is a C-type asteroid, rich in carbon-based materials and less dense than both the S-type silicate rocky asteroids and the more rare M-type iron-nickel asteroids.
Source: Courtesy of NASA and the Applied Physics Laboratory of Johns Hopkins University.

The rubble pile structure can also explain why some asteroids have very low densities. For example, when the 66-kilometer asteroid (253) Mathilde was briefly visited by the NEAR Shoemaker spacecraft in 1997, its mass was determined by noting the deflection of the spacecraft as it flew by. The more massive the asteroid, the more easily its gravity would alter the path of the passing spacecraft. The volume of Mathilde was estimated from a shape model, which was determined from several camera images taken during the spacecraft flyby. Dividing Mathilde's mass by its volume gave a density of 1.3 grams per cubic centimeter. Now, water has a density of 1 gram per cubic centimeter, so if Mathilde were a bit less dense it would float in a bowl of water—albeit a very large bowl of water. This surprisingly low density and an estimated porosity of greater than 60 percent also explain why Mathilde survived several impacts by asteroids large enough to create craters with diameters in excess of Mathilde's radius. These craters are huge and initially it was not clear how a rocky asteroid could withstand such a beating without completely fragmenting into smaller pieces. It was Gene Shoemaker who first pointed out the obvious. The only reason Mathilde survived

FIGURE 6.3. The Japanese spacecraft Hayabusa extensively studied near-Earth asteroid (25143) Itokawa in the fall of 2005. Near-Earth asteroid Itokawa was named for
the father of Japanese rocketry, Hideo Itokawa, and Hayabusa is Japanese for Falcon.
In this image, a very rough surface and the two large lobes that may have come
together long ago to form this object suggest the rubble pile nature of this asteroid.
Itokawa is an S-type silicate rocky asteroid.
Source: Courtesy of JAXA.

such massive-impact cratering events was its extreme porosity. The
substantial void space between Mathilde's solid chunks allowed the
energy of the impacts to be absorbed rather than transmitted throughout the asteroid. If Mathilde were a solid rocky body, it would not have
survived. An analogy would be if a sledgehammer is used to hit a brick,
the brick will shatter into several pieces, but if a large pile of sand is hit
with the same sledgehammer, you'll just get a big indentation. The
porosity of a loose pile of sand is about 40 percent, or 40 percent void
space and 60 percent sand. Mathilde's porosity was estimated at about
60 percent, although whether this porosity is due to large- or small-
scale interior voids is not clear.

Upon its arrival at the near-Earth asteroid (25143) Itokawa in September 2005, the Japanese spacecraft Hayabusa imaged a very rugged
body that looked much like an asteroid that had been subjected to a
catastrophic collision, blasted apart, and reassembled long ago into a
rubble pile structure.

There are two rocky lobes connected by a smooth neck region but
very few impact craters. It seems likely that numerous ongoing

collisions by smaller asteroids shook Itokawa's surface, causing surface materials to adjust. This so-called seismic shaking repeatedly filled in any craters that remained from previous collisions. Spectroscopic investigations into the nature of Itokawa's surface showed that it is made up primarily of the same types of minerals, like olivine and pyroxene, found in ordinary chondrites, the most common type of meteorites. After overcoming spacecraft thruster and communication problems, a loss of spacecraft control gas, and a dead battery, the intrepid Hayabusa spacecraft made it back to Earth in June 2010 and safely deposited its sample capsule containing thousands of Itokawa's surface dust particles on the Australian outback. These dust samples show that Itokawa is a so-called LL chondritic body. It has been extensively heated and its elemental composition is broadly similar to the Sun's initial composition. The "LL" part means that it is low in iron and low in metals generally. These dust particles will be studied for years using comprehensive laboratory techniques. Based on ground-based telescopic studies, MIT's Richard Binzel and his colleagues had suggested that Itokawa's spectrum was similar to LL chondrite meteorites in 2001.

Mathilde's ability to absorb punishment without going to pieces and Itokawa's rugged good looks are evidence of their rubble pile structures, but some of the strongest evidence in support of the rubble pile model of asteroids has come from studies of their rotations.

Whirling Rocks: The Rotation of Asteroids

There is a community of amateur astronomers who devote large portions of their nighttime hours to observing the subtle variations in the light received from distant asteroids. Many of these observers are amateurs in name only. They do extraordinarily precise work to determine the rotation characteristics of asteroids. Imagine an asteroid shaped like a bowling pin—there are a few out there that do resemble a bowling pin. The Sun illuminates our bowling pin–shaped asteroid and this sunlight is reflected to our eye so the object's brightness depends upon the breadth or narrowness of the object presented to the observer. That is, an asteroid is brighter when it presents its broad side to us rather than its pointed end. Unlike a perfectly thrown football that spirals about its longest axis, Mother Nature's preferred rotation state is about an object's shortest axis.

By carefully monitoring the time history of the amount of light received from a rotating asteroid, or its light curve, astronomers can determine its rotation period and often, if the observations are extensive, the likely orientation or direction of the rotation axis in space. The range of asteroid rotation periods is from several weeks to less than 30 seconds. Surprisingly, asteroids larger than about 150 meters in diameter do not, as a rule, have rotation periods shorter than about two hours. Asteroids smaller than 150 meters usually have rotation periods shorter than two hours and most of the relatively fast rotators of substantial size have moons. What's going on here?

The bulk density of rocky S-type asteroids is likely to be about 2.5 grams per cubic centimeter. For example, the density of S-type asteroid Eros, determined from the NEAR Shoemaker spacecraft tracking observations, turned out to be 2.6 grams per cubic centimeter. If we assume these asteroids are solid rock, they should be able to spin at nearly any rate. But if we assume that an asteroid is a nearly spherical, strengthless rubble pile collection of loose rocks, it can only spin as fast as eleven times per day before it would fly apart. So if most asteroids are indeed rubble piles, there should be a rotation barrier of about eleven times per day (or about 2.2 hours per rotation). And there is such a barrier. Voilà, most asteroids are rubble piles. Maybe we've gone a bit too far here since most asteroids are not spherical, most have some strength, and there are a few small, slowly rotating asteroids as well as a few rapidly rotating larger asteroids. But still, the plot shown in figure 6.4 suggests that, in the main, there really is a sort of barrier whereby larger asteroids do not have rotation periods faster than about two hours.

Fast-rotating smaller asteroids, with little strength, would be expected to throw off material from their equatorial regions, which would then likely reassemble into a moon. Figure 6.5 shows a radar image of binary asteroid (66391) 1999 KW4 with an obvious equatorial bulge about its middle. Its rotation period is now about 2.8 hours, but it is likely that it once rotated a bit faster as a result of the spin-up

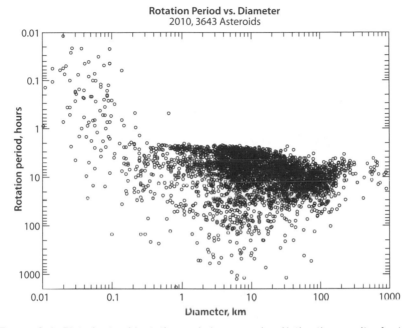

Figure 6.4. Plot of asteroid rotation periods versus size. Notice the paucity of asteroids larger than about 150 meters that have rotation periods less than two hours. Likewise, there are relatively few asteroids smaller than 150 meters that have rotation periods of greater than two hours.

Source: Courtesy of Alan W. Harris.

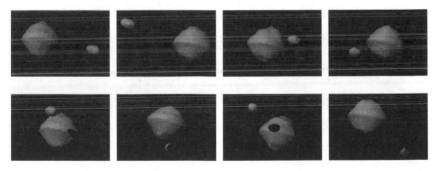

Figure 6.5. Radar observations of near-Earth asteroid (66391) 1999 KW4 taken in May 2001 revealed it to be a binary object with an equatorial bulge on the larger body. The longest dimension of the primary asteroid and its moon are 1.6 and 0.6 kilometers, respectively. The moon orbits the primary asteroid in about 17.4 hours at a distance of 2.5 kilometers.

Source: Courtesy of NASA/JPL-Caltech.

process due to sunlight being re-radiated more effectively from one side of the asteroid with respect to the other—the YORP effect. Then loose surface material rolled downhill from the poles to the equatorial region, flew off the surface, and reassembled into the nearby moon.

At least 15 percent of all near-Earth asteroids have moons; a few have two moons each. In addition, there is a similar fraction that is double lobed, as though they were once asteroids with satellites or double asteroids, but they have since merged.

Radar Images of Near-Earth Asteroids

If a near-Earth object gets close enough to Earth for radar observations from one or both of the two planetary radars located in Arecibo, Puerto Rico, and the Goldstone complex in Southern California's Mojave Desert, the object's shape and rotation characteristics can often be determined to a surprising level of detail. The radar sends a pulse of radio waves toward the asteroid and then measures how long it takes the signal traveling at the speed of light to reach the asteroid, bounce, and then return to the radar-receiving antenna. Since the speed of light is known accurately (nearly three hundred thousand kilometers per second), the distance, often called "range," to the asteroid's nearest bounce point can be determined with an accuracy of a few meters.[3] As the asteroid rotates, different facets of it present themselves to the incoming radar signal and their instantaneous distances can also be determined. For example, a radar signal encountering the small end of a bowling pin–shaped asteroid rotating around its shortest axis would cause a slightly shorter round-trip light time and hence a shorter radial distance than a radar bounce from the broad-side profile of the same asteroid. The radar also allows a radial velocity measurement of the reflecting surface along the line between the radar and the asteroid. This latter measurement is called a Doppler shift because the frequency of the radar signal increases, or shifts, for asteroid facets approaching the radar antenna, remains constant for facets stationary with respect

[3] Nerds often try to get their automobiles to last for one light second or about 186,000 miles.

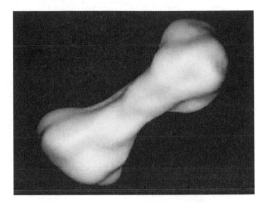

FIGURE 6.6. Radar observations of main-belt asteroid (216) Kleopatra made in November 1999 allowed its dog bone shape to be determined and its surface to be characterized as metallic. Kleopatra is an M-type asteroid and its overall dimensions are 217 × 94 × 81 kilometers.
Source: Courtesy of NASA/JPL-Caltech.

to the radar antenna, and decreases for facets moving away from the radar antenna.

A careful compilation of all the radar signal returns from an asteroid, often in conjunction with analysis of the same object's optical light curve, can be used to create a shape model for the asteroid, discover whether or not it has moons, and determine its rotation period and rotation pole orientation. Radar measurements can also be used to estimate the surface roughness and examine the extent to which the surface is metallic.[1] As seen in figure 6.6, radar observations of main-belt asteroid (216) Kleopatra reveal it to be a dumbbell-shaped object made up of a loose mixture of metal and rock over two hundred kilometers in extent in its longest dimension. In September 2008, Pascal Deschamps of the Paris Observatory, Franck Marchis from the University of California at Berkeley, and their colleagues used the giant Keck II ten-meter aperture telescope atop Hawaii's Mauna Kea summit to determine that

[1] Scott Hudson at Washington State University first provided the sophisticated computer techniques to estimate an asteroid shape model from a set of radar range and Doppler measurements. Some of the other pioneers in this relatively new field of asteroid shape modeling include Mike Nolan at Arecibo, Don Campbell at Cornell University, Jean-Luc Margot and Michael Busch at UCLA, and Lance Benner, Marina Brozovic, and especially the late Steve Ostro at JPL.

TABLE 6.1. Spacecraft observations of asteroids

For each spacecraft encounter with an asteroid, the best pixel scale, encounter time, and encounter distance are given.

Target Body	Best Pixel Scale (meters)	Mission Name	Encounter Arrival or Flyby Date	Distance (kilometers)
Gaspra	54	Galileo[1]	10/29/1991	1,600
Ida/Dactyl	30	Galileo	08/28/1993	2,391
Mathilde	160	NEAR[2]	06/27/1997	1,212
Eros	0.05	NEAR	02/12/2000	Rendezvous & Landing
Braille[3]	120	DS1	07/29/1999	28
Annefrank[4]	185	Stardust	11/02/2002	3,079
Steins	80	Rosetta[5]	09/05/2008	803
Lutetia	60	Rosetta	07/10/2010	3,162
Itokawa	0.06	Hayabusa[6]	09/2005	Rendezvous & Landing
Vesta	20	DAWN[7]	07/2011	Rendezvous
Ceres	70	DAWN	02/2015	Rendezvous

[1] The Galileo spacecraft flew past asteroid (951) Gaspra and (243) Ida. The longest dimensions of Gaspra and Ida are 19 and 54 km, respectively, and a 1.6-km-sized moon (Dactyl) was discovered orbiting Ida. Dactyl was the first of many asteroid moons to be discovered.

[2] The Near-Earth Asteroid Rendezvous (NEAR) mission discovered four large craters on (253) Mathilde that are comparable in diameter to the asteroid's radius. The NEAR Shoemaker spacecraft spent about a year in orbit about near-Earth object (433) Eros before the spacecraft was boldly landed on the asteroid's surface. Mathilde is roughly spherical with a longest dimension of 66 km while Eros is shaped a bit like a fat sausage with a longest dimension of 34 km.

[3] (9969) Braille, about 2 km in its longest dimension, has a long rotation period of 226 hours. While the close approach distance was only 28 km, this object was imaged at a distance of 13,500 km by the Deep Space 1 spacecraft.

[4] Asteroid (5535) Annefrank has a roughly ellipsoidal shape with a longest dimension of 6.6 km.

[5] The European Space Agency's Rosetta spacecraft, on its journey to comet Churyumov-Gerasimenko, flew past asteroids (2867) Steins and (21) Lutetia. Steins is about 6.7 km in its longest dimension and was imaged with both the narrow-angle and wide-angle cameras, but a shutter problem on the narrow-angle camera prevented images from that instrument near its closest approach. Lutetia's long dimension is about 130 km.

[6] On two occasions in November 2005, the Hayabusa spacecraft touched down on the surface of near-Earth asteroid (25143) Itokawa. Although the spacecraft sampling mechanism failed during each touchdown attempt, dust particles stirred up by contact with the surface were captured and brought back to Earth on June 13, 2010.

[7] The DAWN spacecraft carried out a rendezvous with asteroid (4) Vesta in July 2011, and there are plans to carry out a rendezvous with dwarf planet Ceres in 2015.

Kleopatra had not one but two moons with diameters of about five and three kilometers. These outer and inner moons have since been named Alexhelios and Cleoselene after Cleopatra's twin children, Alexander Helios and Cleopatra Selene II, both fathered by Mark Anthony.

The finest asteroid shape models are enabled by spacecraft imaging. Table 6.1 outlines the spacecraft encounter dates, the asteroid encounter distances, and the resolution or "pixel scales" for the various spacecraft missions. A surface feature can be seen, or resolved, if it extends over a few image pixels. For example, the best pixel scale for asteroid (951) Gaspra was 54 meters, so an object as small as about 150 or 200 meters on the surface of Gaspra could be seen in the Galileo spacecraft images.

Comets: The Minority Members of the Near-Earth Object Congress

In terms of numbers, active comets represent a minority of only about 1 percent in the near-Earth object population with asteroids making up the remaining 99 percent. Active comets are usually separated into two different orbital classes, the short-period comets whose motions are most often controlled by the gravitational tugs of Jupiter and the long-period comets that arrive in the inner solar system directly from the distant Oort cloud.

Whereas asteroid structures are dominated by collisional events, cometary structures seem to be dominated by the vaporization and loss of their ices and the small particles that are embedded within. Asteroids can be blasted apart by collisions over millions of years whereas the far more fragile comets last only a fraction of this time before either exhausting their volatile ices or having their ices buried and insulated by overlying layers of dust. In either case, they become inactive and hence would then be called asteroids. That is apparently the only defining difference between comets and some asteroids. Comets have activity in the form of gas and dust loss while asteroids do not. It is estimated that perhaps 15 percent of all near-Earth asteroids are either ex-comets or dormant comets. End states, or sinks, for comets can

FIGURE 6.7. Hubble Space Telescope image of disintegrating comet 73P
Schwassmann-Wachmann 3 taken on April 18, 2006.
Source: Courtesy of NASA, European Space Agency; H. Weaver (Applied
Physics Laboratory of Johns Hopkins University), M. Mutchler, and Z. Levay,
Space Telescope Science Institute.

include this transformation into an asteroid. A comet's life can also
end with a dramatic series of cascading splitting events often followed
by a complete disintegration into a cloud of dust. The cometary disintegration process was observed by the Hubble Space Telescope in 2006
when short-period comet 73P/Schwassmann-Wachmann 3 underwent
a cascading series of nucleus-splitting events. There is a few percent
chance that any given comet will split upon entering the inner solar
system, but apart from a few examples where a comet has passed close
to Jupiter or the Sun and undergone splitting due to the tidal effects,
the mechanism as to why some comets split is not well understood.[5]

[5] About three dozen comets have been observed to split to date with four of those being due to
tidal effects. For example, after comet D/Shoemaker-Levy 9 was captured into an orbit around
Jupiter in about 1929, it closely approached Jupiter's surface in July 1992 to within a distance
only one-third the planet's radius. At that time, Jupiter's gravitational pull on the comet's closest
edge slightly exceeded Jupiter's pull on the comet's most distant edge so the comet experienced
a slight stress that pulled its fragile structure apart. Each of the twenty-plus fragments then
plowed into Jupiter's atmosphere two years later in July 1994 with relative velocities of about

But split they do, so cometary nuclei must be very fragile; were you to hold a lump of a comet's nucleus in your hand, you could easily break it apart like a loosely packed snowball.

The solid portion of comets, their nuclei, are more difficult to study than asteroids because when the nuclei enter the inner solar system and are close enough for Earth-based telescopic observations, the vaporization of their ices and the release of the embedded dust particles forms a coma or atmosphere that completely obscures the central nucleus. When this atmosphere finally subsides as the comet retreats from the inner solar system, the bare nucleus is generally too far from the Earth to be easily studied. As a result, much of what we know about cometary nuclei has come from the few spacecraft missions that have flown within the comet's atmosphere to get a better look at the nucleus. Table 6.2 outlines the various missions that have examined the cometary phenomena up close. While each of these cometary nuclei is different in appearance, some general conclusions can be drawn. Most of the nuclei are made up of dust particles and clumps embedded in, or overlaying, volatile ices.

The most prevalent ice in comets, by far, is water ice. Carbon dioxide ice, or dry ice, is often present at the 10 percent level or less although the carbon dioxide abundance levels were considerably higher for the very active comet 103P/Hartley 2. On November 4, 2010, a spacecraft flyby of this comet revealed a bizarre peanut-shaped nucleus where dust and water ice clumps were blown off the small end by carbon dioxide gas jets while only water vapor was released from the smooth neck region. This comet must have spent only a limited time in the inner solar system since, were this not the case, most of the volatile carbon dioxide would have been warmed by the Sun and vaporized long ago. Often other even more volatile ices like carbon monoxide,

---

sixty kilometers per second. This comet and a few others now carry a "D" prefix to indicate that they no longer exist or are deceased. Another comet, 16P/Brooks 2, split during a close approach to Jupiter in 1886 and two more comets split after approaching the sun's surface to within two solar radii. The tidal forces are very gentle so these comets, and perhaps others as well, are very fragile structures. For the majority of the split comets that were not caused by tidal effects, it is not clear what triggered their fragmentation, but disruption as a result of rapid rotation is one possibility.

TABLE 6.2. Spacecraft observations of comets

For each spacecraft encounter with a comet, the best pixel scale, encounter time, and encounter distance are given.

Target Body	Best Pixel Scale (meters)	Mission Name	Arrival or Flyby Date	Encounter Distance (kilometers)
Giacobini-Zinner		ICE[1]	9/11/1985	7,800
Halley		VEGA 1[2]	03/06/1986	8,890
		Suisei[3]	03/08/1986	150,000
		VEGA 2	03/09/1986	8,030
		Sakigake	03/11/1986	7 million
	45	Giotto[4]	03/14/1986	596
Borrelly	47	DS1[5]	09/22/2001	2,171
Wild 2	15	Stardust[6]	01/02/2004	240
Tempel 1	1	Deep Impact[7]	07/04/2005	500
Hartley 2[8]	4	Deep Impact	11/04/2010	700
Tempel 1[9]	10	Stardust-NExT	02/15/2011	178
Churyumov-Gerasimenko		Rosetta[10]	mid-2014	

[1] The International Cometary Explorer (ICE) spacecraft passed 21P/Giacobini-Zinner on the side away from the sun about 7,800 km from its nucleus. It carried no camera but did detect the magnetic field of the sun wrapped around the comet's ion atmosphere. Earlier in its career, the ICE spacecraft was used to study the interaction of the solar wind particles with the Earth's charged particle and magnetic field environment from a point between the Sun and Earth. After four years of these observations, a series of five lunar gravity assists designed by Robert Farquhar were carried out and the International-Sun-Earth-Explorer 3 (ISEE-3) spacecraft was then retargeted (and renamed ICE) for a flyby of comet Giacobini-Zinner.

[2] The Soviet Union sent two spacecraft to comet 1P/Halley. The camera on the VEGA 1 space-craft was somewhat out of focus and the camera on the VEGA 2 spacecraft obtained overexposed images of the comet's nucleus. Comet Halley's longest dimension is about 15 km.

[3] Japan's Suisei and Sakigake spacecraft were designed to study the solar wind interaction with the comet's atmosphere. Neither spacecraft carried a camera.

[4] The European Space Agency's Giotto spacecraft was programmed to image the brightest object in its field of view, so the highest-resolution nucleus images were located adjacent to a bright dust jet.

[5] The Deep Space 1 (DS1) spacecraft, designed to test new technologies including an ion drive engine, flew past comet 19P/Borrelly and used its camera to provide images of an elongated bowling pin–shaped nucleus whose longest dimension was about 8.4 km. The highest resolution image was taken at a distance of 3,556 km.

[6] The Stardust spacecraft collected cometary dust samples during the flyby of 81P/Wild 2 and returned them to Earth for study on January 15, 2006. The longest axis of this roughly spherical nucleus is about 5.5 km. The main Stardust spacecraft, now renamed Stardust-NExT, was retargeted for a flyby of comet 9P/Tempel 1 on February 15, 2011. "NExT" was added to the mission name to denote "New Exploration of Tempel 1."

(continued)

TABLE 6.2. *Continued*

[7] Comet 9P/Tempel 1 was observed from the flyby spacecraft from a distance of about 700 km. The spacecraft then stopped imaging and switched to "shield mode" to mitigate near-nucleus dust hits. The impactor spacecraft was able to take high-resolution images and radio them back to the flyby spacecraft before vaporizing during the impact itself. This comet's nucleus is sort of a rounded pyramid with a longest dimension of 7.6 km.

[8] After its flyby of comet Tempel 1 in July 2005, the Deep Impact spacecraft was retargeted for a flyby of comet Hartley 2 in November 2010. The nucleus of comet 103P/Hartley 2 was a bizarre sight to see. It is shaped like an elongated peanut, 2.3 km in its longest dimension and with a smooth neck connecting two larger end members that were very rough.

[9] The nucleus of comet 9P/Tempel 1 was observed in July 2005 by the Deep Impact spacecraft and again in February 2011 by the Stardust-NExT spacecraft. The February 2011 observations identified the impact crater created by the 2005 Deep Impact mission. The crater was about 50 meters in diameter but quite indistinct and subdued with a raised central mound, suggesting that the nucleus has a fragile and porous surface.

[10] The European Space Agency's Rosetta mission will rendezvous with comet 67P/Churyumov Gerasimenko in mid-2014, observe the active nucleus for several weeks through perihelion, and deploy a lander to closely investigate the surface of the nucleus. If successful, the nucleus lander could achieve a pixel scale of 2 centimeters per pixel.

methane, and ammonia are also present in comets but at much reduced levels. Since water is nearly ubiquitous in the outer solar system where comets formed and is the least volatile of the ices mentioned, it is not too surprising that water ice dominates the icy portion of cometary nuclei. The evidence so far suggests that these ices are buried beneath the surface dust layers of comets. If that were not the case, and the ices were exposed on the surface, they would not last very long in the relatively warm inner solar system. Before the nucleus of comet 1P/Halley was visited by spacecraft in 1986, cometary nuclei were thought to be dirty ice balls or dirty snowballs. However, the paucity of ices on the nucleus surface and the realization that the non-volatile particles were more abundant than the ices led to a slight revision in terminology. The dirty ice balls became icy dirtballs.

When an active comet enters the inner solar system, the Sun's heat warms the comet's dusty surface and the ices just below this dust layer begin to vaporize. Because there is no pressure to speak of, the vaporization of these ices generates gases without first going through the liquid phase, a process called sublimation. The vaporization of these ices

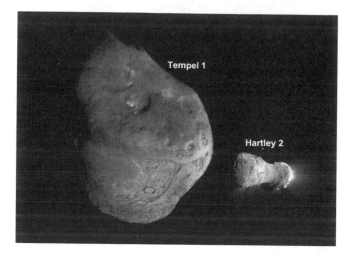

FIGURE 6.8. The nuclei of 9P/Tempel 1 and 103P/Hartley 2 were imaged by the Deep Impact spacecraft on July 4, 2005, and November 4, 2010, respectively. The nucleus of Tempel 1 has an approximate diameter of 6 kilometers and has predominantly water vapor outgassing with a lesser amount of carbon dioxide emission. The peanut-shaped nucleus of Hartley 2 is just over 2 kilometers in its longest dimension and has carbon dioxide gas jets that lift dust and water ice clumps off the ends of the nucleus while only water vapor was detected emanating from the smooth neck region. *Source*: Courtesy of NASA and the University of Maryland.

then drives gas and dust jets that have been obvious from both ground-based observatories and nearby spacecraft.

Apart from a few obvious ancient craters on the surface of comet 9P/ Tempel 1, the cometary nuclei observed by spacecraft are nearly devoid of the impact craters that are so prevalent on the surfaces of many asteroids. Since comets also have collisional events, their craters must be eroded away rather quickly by surface activity. Some crude estimates of the amount of material lost by a short-period comet during each passage through perihelion suggests that one to two meters of surface material is lost at each return to the Sun's neighborhood. For a cometary nucleus of a few kilometers in extent, this mass loss allows the active comet to exist for only about one thousand returns before it runs out of the ices that drive the cometary activity. Compared to their asteroid brethren, active comets are short-timers in the inner solar system.

Transition Objects: Ambiguous Asteroids and Closet Comets

Not so long ago, comets were thought of as dirty ice balls traveling in highly inclined and eccentric orbits about the Sun, and the vast majority of the rocky asteroids traveled in nearly circular, highly flattened orbits within the asteroid belt between the orbits of Mars and Jupiter. Comets traveled in either direction around the Sun (prograde or retrograde orbits) while the asteroids, like the planets, traveled only in prograde orbits about the Sun. But today there are so many exceptions to these perceptions that one must conclude that the line between comets and asteroids is no longer clearly drawn. Today we have asteroids that travel about the Sun in elongated, highly inclined, even retrograde orbits (e.g., 20461 Dioretsa). Perhaps these are ex-comets. There are active comets on flattened, nearly circular orbits within the asteroid main belt, and there are more than a few comets that may be evolving from active comets to inactive asteroids at the current time. In this latter group are objects that at one time appeared as active comets but now spend most or all of their time as inactive asteroids. These transition objects, which carry both an asteroid and a comet designation, include (4015) Wilson-Harrington, (7968) Elst-Pizarro, (118401) Linear, (60558) Echeclus, (2005 U1) Read, and (2008 R1) Garradd. Another dual designation object, (2060) Chiron, may be a centaur object transitioning to an active comet.

Just to confuse the issue further, consider "comet" P/2010 A2, which was discovered by the LINEAR sky survey on January 6, 2010. It is termed a comet since it shows activity and it is in the inner region of the main asteroid belt; initial thinking suggested it might be one of the few main-belt comets. But the atmosphere showed only dust and no gases. Further inspection by UCLA astronomer David Jewitt using the Hubble Space Telescope on various dates from January 29 through May 29, 2010, revealed a bizarre dust atmosphere with the "nucleus" on one end of an X-shaped filament structure. Since normal comets have their nuclei embedded within their dust and gas atmospheres, this object is now considered to be the result of two asteroids colliding a year earlier. So now we have a battered asteroid falsely identified with

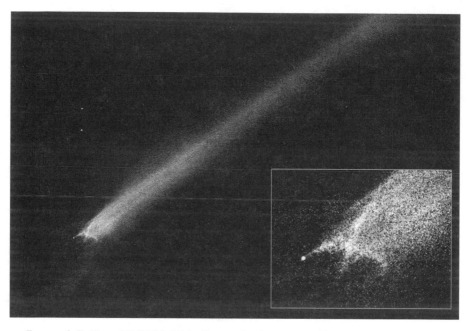

Figure 6.9. Comet P/2010 A2 is the result of two asteroids colliding in the inner region of the main asteroid belt. The largest fragment of this collision process is seen at the very end of the X-shaped dust filaments that resulted from the impact several months earlier. The Hubble Space Telescope Wide Field Camera 3 took this image on January 29, 2010.
Source: Courtesy of NASA, European Space Agency, and D. Jewitt (University of California, Los Angeles).

a cometary designation, further underscoring the need to redefine the comet and asteroid classes.

Summary

Active comets seem to be composed of a weak, dusty, silicate matrix material that is in part enriched with ices—mostly water ice. On the nucleus surfaces the ices have been depleted, leaving a weak matrix of non-volatile material that can form large fragile structures that would seem to have difficulty surviving for more than a few dozen returns to the Sun's neighborhood. Comets can lose their ices and split into cascading pieces or disintegrate entirely into a cloud of dust. Some survive as fragile inactive objects that form end members of the very diverse asteroid population.

Asteroids have on one end of their spectrum of structures the fragile ex-comets. Most asteroids would seem to be collections of collision fragments or rubble piles (e.g., Mathilde, Itokawa), shattered rocks (Eros, Gaspra, Ida), or slabs of solid nickel-iron. Some asteroids have water resources in the form of hydrated minerals.

Donald Duck and Uncle Scrooge McDuck made prescient "observations" in 1960, noting that asteroids were of diverse shapes, had moons, and were largely composed of loose structures that resembled rubble piles. However, they were only looking for a safe asteroid storehouse to hide Uncle Scrooge's vast riches. What they failed to realize was that with their minerals, metals, and water resources, asteroids themselves are vast storehouses of riches.

CHAPTER 7

Nature's Natural Resources and the Human Exploration of Our Solar System

All exploration is human exploration.

Why Should We Explore Near-Earth Objects?

The reasons for exploring and studying near-Earth objects go well beyond intellectual curiosity. These objects represent the least changed and hence most primitive objects within our solar system and, as such, they provide critical clues as to the origin of our solar system. If we wish to understand the chemical mix and thermal environment from which our solar system arose 4.6 billion years ago, then the compositions of near-Earth objects, and the meteorites derived from them, will offer insights into the conditions present when the planets formed. Armed with these clues as to the formation conditions and an albeit incomplete understanding of the present conditions within the solar system, scientists can provide reasonable models for the 4.6-billion-year evolutionary path between then and now (see chapter 3).

The study of near-Earth objects plays a key role in understanding the mechanisms that brought the building blocks of life to the early Earth. The Earth formed hot without significant supplies of water and organic materials so once the Earth cooled down, near-Earth objects likely delivered much of these materials to the early Earth (see chapter 4).

In addition to their importance in studying the conditions under which our solar system formed and the mechanism by which the early

Earth received a veneer of water and organic carbon-based materials that allowed life to form, near-Earth object collisions with Earth have the capacity to cause catastrophic damage to Earth and all its complex life forms. Although the likelihood of a large near-Earth object impacting Earth, causing the type of extinction inflicted upon the dinosaurs sixty-five million years ago, is extremely remote, we need to be vigilant. We'll need to find them before they find us. The hazards posed by near-Earth object collisions will be addressed in chapter 8. Fortunately, a collision by a near-Earth object is a natural disaster that we can prevent if we find them early enough (see chapters 9 and 10).

This chapter will address two additional reasons for investigating and exploring near-Earth objects.

1. *Mining*: Some near-Earth objects are rich in minerals that are extremely expensive and relatively rare on the Earth's surface. In addition, future structures and habitats in Earth orbit or in near-Earth space would benefit from the use of raw materials already in space. One day the mining of near-Earth asteroids could be the backbone of a new space industry.

2. *Human exploration*: If the planet Mars and its two moons are the ultimate destination for human exploration beyond the Moon, then the much more easily accessible near-Earth objects could provide a technological precursor mission to the much more difficult human exploration of Mars. A near-Earth object could provide a convenient stepping stone to the human exploration of Mars.

Mining Near-Earth Objects

The interest in mining near-Earth asteroids is the result of the paucity of some critically important rare metals on the Earth's surface and their relative abundance on near-Earth asteroids. During the Earth's formation process, most of its heavy metals like iron and nickel and especially the platinum group metals (platinum, palladium, rhodium, iridium, osmium, and ruthenium) sank to the Earth's core, leaving the Earth's upper crust relatively metal poor.

Because of its purity, stability at high temperatures, and extreme resistance to corrosion, platinum is utilized in a variety of industrial processes, in circuitry, in crafting fine jewelry, and in the automobile catalytic converters that treat exhaust to reduce pollutants. The market price of platinum is typically higher than gold because it is rare and attractive with far more uses than gold.[1] Far less than 1 percent of the Earth's supply of the platinum group metals is in the Earth's crust and accessible. More than 70 percent of the supply that is accessible is located in the Bushveld igneous complex in South Africa where molten rock from the Earth's mantle was brought to the surface through long vertical cracks in the Earth's crust. These relatively rich igneous ores have a platinum group metals concentration of 10 parts per million (ppm) or about 10 grams per metric ton. From meteorite evidence, some asteroids have ten times that concentration of platinum group metals, or 100 ppm.

Although there are asteroids that are nearly solid nickel-iron, silicate rocky asteroids are far more common in near-Earth space and contain up to 20 percent metals in the form of small particles dispersed throughout the silicate rock. Furthermore, it would be easier to crush the surfaces of these rocky asteroids and extract the metals than it would be to mine solid metal asteroids. For an example, let's assume that an easily reached, rocky near-Earth asteroid has a diameter of one kilometer, a bulk density of 2.5 grams per cubic centimeter, and a 20 percent metal content with that metal ore having a platinum concentration of 100 ppm. This asteroid would then contain 26,000 tons of platinum. With the typical price of platinum of $1,700 per ounce, or $60 per gram, the value of the platinum alone in this asteroid would be $1.6 trillion. The market value of the nickel and iron in this body

---

[1] The relative value of platinum with respect to gold is even acknowledged in the credit card industry and airline frequent flier clubs where a platinum card is considered more prestigious than a mere gold card. As yet the very rare metal rhodium, a highly reflective and corrosion resistant metal used in jewelry, mirrors, and catalytic converters and alloyed with platinum for aircraft turbine engines, is even more expensive than platinum. However, rhodium's superior value has not yet been acknowledged with its own credit card.

would be another $3.9 trillion. Our kilometer-sized near-Earth asteroid contains more precious metals than have been mined on Earth in the entire course of human history.

Mining near-Earth asteroids for their precious metal content would make sense if it were cost-effective to retrieve these resources and bring them back for use on Earth. However, there would be significant infrastructure costs involved in space mining, and the launch costs alone to emplace the mining equipment would be enormous. The current launch costs to place one kilogram of stuff into low-Earth orbit (about three hundred kilometers above the Earth's surface) is roughly $10,000 per kilogram and it would be more expensive to reach the asteroid itself. However, as the Earth's resources of precious metals are depleted and launch costs decrease with time to under $1,000 per kilogram, returning precious metals from these objects could make economic sense.[2]

In the next few decades, mining near-Earth asteroids for building structures and rocket supply depots in near-Earth space might make more economic sense than actually returning precious metals to Earth. In the near future when space tourism and Earth-orbiting hotels take hold and orbiting solar energy mirrors are used to collect sunlight and microwave the collected energy down for use on the Earth's surface, the cost-effectiveness of near-Earth object resources will begin to look more attractive.[3] Their metals could be mined and processed to build structures. Some near-Earth asteroids have hydrated minerals, or clays, so their ores could be processed for their water content. Water would be required to sustain life and could be used to make an effective radiation shield against cosmic rays, which are high-speed charged particles (mostly protons) that can cause damage to life forms. The water, in

[2] An isotope of helium, He3, might one day be the most valuable space resource since it may be used in fusion reactors with the hydrogen isotope deuterium to provide an attractive source of energy. He3, implanted by the solar wind over millions of years on the ancient lunar surface, could be collected, whereas the Earth's atmosphere prevents this light form of helium from reaching the Earth's surface.

[3] The collection of sunlight in high Earth orbit would be far more efficient than collectors on the Earth's surface because the atmosphere reflects about 30 percent of the sun's light and collectors operating at a geosynchronous altitude of about 36,000 kilometers could operate nearly continuously and send this energy down to a specific collecting area on the Earth's surface. These benefits would have to be weighed against the downsides to beaming the energy through the atmosphere.

turn, could be broken down by electrolysis into hydrogen and oxygen, the most efficient form of rocket fuel, so that water and fuel depots in Earth orbit could be used to resupply rocket cargo ships during the near-Earth commercialization that is sure to come. Near-Earth objects may one day become the fueling stations and watering holes for future interplanetary exploration.

Human exploration is usually driven not by the quest for knowledge but for commercial gain. Just as Columbus is often credited with finding the Americas as a result of his desire to find a shorter trade route to the Indies, it may be that in-orbit and near-Earth space capitalism, with its corporations and stockholders, will one day drive the exploration of near-Earth objects.[4]

The Human Exploration of Near-Earth Objects

While the commercial exploitation of near-Earth objects may one day be a viable, perhaps even necessary, activity, there are more short-term reasons to explore Earth's closest neighbors—including using them as stepping-stones to Mars. On April 15, 2010, President Obama outlined a U.S. National Space Vision that included human exploration of a near-Earth asteroid by 2025 as a first step toward human exploration of the planet Mars. Assuming that human exploration of Mars and its two moons is the overarching goal of the American space program, a first step toward reaching that goal would include a target that is both easier to reach and interesting in its own right. Near-Earth asteroids fill both bills.

In terms of landing upon and returning from a celestial body, traveling to and from some near-Earth asteroids would require less fuel than a round-trip to the Moon. Space mission planners often use the required change in spacecraft velocity (delta V) to measure the effort, or fuel, required to get from point A to point B. For example, if a spacecraft on

---

[4] The current space treaties, signed by most of the space-faring nations, prohibit signatories from claiming sovereignty of celestial bodies, but private ownership by private companies may not be excluded.

the Earth's surface can be given a velocity of 11.2 kilometers per second (km/s), it can completely escape the Earth's gravity. So 11.2 km/s is termed the Earth's escape velocity. It takes 8.0 km/s to reach low Earth orbit (LEO) at an altitude of about 300 km and hence another 3.2 km/s to escape the Earth from LEO. From LEO to the Moon's surface requires about 6.3 km/s whereas from LEO to the surface of some near-Earth asteroids would be less at about 5.5 km/s. The advantage of landing upon and returning to the Earth from a near-Earth asteroid, rather than the Moon, is even more significant since the Moon's gravity is 47,000 times stronger than that of a 100-meter-sized asteroid, so a good deal less fuel is required to land upon and escape from the asteroid. In addition, the space mining of metals on near-Earth asteroids would be more viable than on the Moon, since the metal concentrations on the surfaces of some near-Earth asteroids are a few hundred times the corresponding lunar concentrations.

While it takes longer to travel to and from a near-Earth asteroid than the week and a half to and from the Moon, a near-Earth asteroid round-trip can be carried out within six months whereas a round-trip to Mars takes more than two years.[5] A short human exploration of a near-Earth asteroid would offer crucial operational experience for the much more difficult round-trip to Mars. Tests could be carried out to better understand how astronauts would move about the surface of an object that has very little gravity and an escape velocity of only 6 centimeters per second for a 100-meter-sized object. Any attempt to walk on the asteroid's surface would be frustrated by putting oneself into orbit with the slightest foot movement. Could astronauts lasso the asteroid or affix cargo nets to the surface so they could "walk" hand

[5] Round-trip times to near-Earth asteroids would depend upon a number of factors, including how similar the asteroid's orbit is to that of the Earth, the stay time at the asteroid, how many large launch vehicles were available, and whether or not fuel and supplies had been cached in Earth orbit so the astronaut's spacecraft could be mated with a cargo vehicle in Earth orbit before heading out to the target asteroid. The easiest asteroids to reach and return from are those that have orbits similar to Earth's. That is, they have nearly circular orbits in the same orbital plane and at about the same distance from the sun as the Earth itself. These are often Aten class near-Earth objects. Asteroids in these types of orbits are the easiest to exploit in terms of their mineral and metal wealth. As we will see in chapter 9, these asteroids are also the most hazardous as impact threats so the same group of objects represents both the most valuable and the most dangerous objects in the solar system.

FIGURE 7.1. An artist's conception of an astronaut
hanging onto a line secured to a near-Earth asteroid.
Source: NASA artist.

over hand around the object? Could a penetrating harpoon or some
sort of epoxy glue be used to provide the necessary handrails? Would
the astronaut space suits even need provisions for leg movement? What
sort of radiation environment would the astronauts face and could they
get along without strangling each other after spending several months
in very close proximity? These are the types of questions that a human
exploration of a near-Earth asteroid could address.

The human exploration of near-Earth asteroids could test the water
recovery systems that would be necessary for the longer flights to Mars.
Water is necessary for astronaut health, and because of its high hydrogen
content, it is also an excellent shield for the cosmic rays and solar proton
events that could cause serious tissue damage to our intrepid near-Earth
object explorers.[6] A water jacket surrounding the crew quarters could

---

[6] In general, men are less susceptible to radiation tissue damage than women and older men are
less susceptible than younger men, so perhaps the ideal crew members for the human exploration of near-Earth asteroids and Mars are geezer astronauts. Old guys rule!

help make a valuable life-sustaining shield and, at the same time, provide the water resources necessary for drinking and hygiene.

Any human exploration of a near-Earth asteroid would likely be preceded by at least one robotic spacecraft rendezvous to determine the object's size, shape, mass, density, structure, rotation characteristics and chemical composition, as well as whether there were any hazardous surface features. The robotic spacecraft would also look for asteroid satellites and make note of the nature of the dust and material on the surface. One of the biggest problems for human exploration is likely to be the dust particles that may be covering the asteroid's surface. A robotic spacecraft could disturb the surface dust and observe the electrostatic stickiness (static cling) of the dust. How likely is the dust to stick to an astronaut's space suit and helmet visor, and how badly will this gritty material affect delicate instrumentation and space suit joints? Would asteroid astronauts soon resemble Pig Pen, the dirt-covered member of Charlie Brown's gang in the Peanuts cartoon strip? A robotic spacecraft could test the dust environment while at the asteroid or bring a sample back to Earth for study.

One of the main scientific goals of the subsequent human exploration mission would be to collect surface samples so they could be extensively studied in Earth-based laboratories. Are the organic surface materials and water similar to those on Earth such that an assessment could be made as to the likelihood that asteroid collisions provided the early Earth with the building blocks of life? How rich is the soil in precious metals? What percentage of the soil could be turned into water resources? An examination into the asteroid's structure would be helpful if another asteroid of this type were to be discovered on an Earth-threatening trajectory. The inclusion of astronauts on a mission to a near-Earth asteroid would greatly improve the quality of the sample collection process and the determination of the object's physical characteristics. They could measure the bearing and sheer strength of the asteroid's surface along with its penetrability and thermal conductivity. Astronauts would also offer an immediate adaptability to unanticipated situations and unexpected opportunities while examining the asteroid's surface features.

The benefits of human exploration of a near-Earth asteroid should include experiments and sample collections that would shed light upon the origin of the solar system and life itself, assay the mineral and metal wealth of the object, and provide information that might one day be useful if we need to mount a spacecraft mission to deflect an Earth-threatening object. However, the main goals of human exploration of a near-Earth asteroid would be to examine possible problems and test some of the procedures that would be required for the longer human exploration of Mars. The astronauts would likely test different techniques for moving about the asteroid's surface, test mechanisms that would extract water from hydrated asteroid minerals and astronaut waste products, and test the effectiveness of radiation shielding materials including a water jacket surrounding the crew cabin. In addition, the near-Earth object mission would likely provide guidelines for the longer Mars mission in terms of the crew's muscle maintenance and state of mind when weightlessness is the norm, mother Earth is just a distant bluish dot, hygiene is difficult to maintain, and personal privacy is just not possible.

Despite the difficulties involved with human spaceflight, it is crucial that we resume these efforts after our more than forty-year hiatus. The scientific and commercial reasons are clear, but even more compelling is the need to transplant self-sustaining human communities on celestial bodies besides Earth to maintain the species against catastrophe—either celestial or self-inflicted.

✧ CHAPTER 8 ✧

Near-Earth Objects as Threats to Earth

> What happens if a big asteroid hits Earth? Judging from
> realistic simulations involving a sledgehammer and a common
> laboratory frog, we can assume it will be pretty bad.
> —Dave Barry

A Hard Rain Is Falling

More than one hundred tons of near-Earth object material pummels
the Earth daily. Fortunately, the vast majority of this material is dust
and pebbles that are too small to survive passage through our thin
atmosphere.[1] Just as hammer blows to a brick and the resulting fragments produce far more small particles than large ones, the continued
collisions of near-Earth objects with other asteroids in interplanetary
space over millions of years produces far more small ones than large
ones. For example, there are thought to be about a thousand near-
Earth objects larger than one kilometer in diameter but more than one
million near-Earth objects larger than thirty meters.

Many of the interplanetary dust and sand-sized particles that rain
down upon the Earth daily result in the harmless meteors, or shooting
stars, that are so enjoyable to watch in the night sky. Occasionally

---

[1] If the Earth were reduced in size to that of an apple, the thickness of almost our entire atmosphere would be that of the apple's skin. It is only this thin, fragile atmosphere that keeps Earth
habitable.

there are meteor showers, or storms, when the Earth passes through the debris thrown off by passing comets and less frequently by asteroids. Notable annual meteor showers are the Perseids, Leonids, and Geminids that take place in August, November, and December when the Earth passes through the debris of comet Swift-Tuttle (Perseids), comet Tempel-Tuttle (Leonids), and asteroid 3200 Phaethon (Geminids).

While the far more numerous particles that make up the small end of the near-Earth object population are the ones that most frequently strike the Earth's atmosphere, it is the rarer but larger objects that can cause serious damage. The vast majority of these objects are stony, and as a general rule stony near-Earth objects smaller than about thirty meters are unlikely to cause significant ground damage. They can, however, create impressive fireball events. Stony objects between thirty and one hundred meters will most likely not impact the Earth's surface but they will result in air blasts that can cause significant damage at the surface. Stony objects larger than about one hundred meters will usually punch through the atmosphere and impact the ground or, more likely, the oceans since Earth's surface is about 70 percent water.

Atmospheric Entry, Breakup, and Air Blasts

The average near-Earth asteroid would enter the Earth's atmosphere at a typical velocity of about 17 kilometers per second. Beginning at an altitude of about 100 kilometers, an Earth-colliding stony asteroid will begin to feel the resistance, or drag, of the atmosphere. As the air resistance increases, the pressure on the front side exceeds the pressure on the backside and the object begins to flatten, or pancake. This flattening process increases the lateral dimensions of the impactor and thus further increases the air resistance. The stress will then cause the object to fragment. Like the parent body, fragments are strongly heated by the air resistance and some of their material ablates away, efficiently carrying heat away from the remaining fragments. For this reason, small fragments of asteroids that penetrate the Earth's atmosphere and land, called meteorites, often hit with little more heat than the ambient temperature of the landing area. Meteorites do not land red-hot and start fires.[2] This same process of heat removal by ablation is used in heat shields that protect spacecraft that are reentering the Earth's atmosphere. These heat shields are designed to lose their red-hot outer layers of material in an effort to keep the capsule cool as it enters the Earth's atmosphere.

If the atmospheric drag is great enough, the aerodynamic pressure will exceed the strength of the impacting asteroid material. The asteroid will then disintegrate and its kinetic energy will be deposited into a relatively small volume of air. This air is strongly heated and expands very quickly to produce an explosion or airburst. The airburst then generates a powerful blast wave consisting of an abrupt pressure pulse followed by a substantial wind. An air blast can do far more ground damage than an object with the same energy that actually reaches the ground. That's why military bombs are designed to burst above ground rather than explode upon impact. For example, an explosion with an equivalent energy of one million tons of TNT (one megaton) would do more than twice the ground damage if it explodes one kilometer above the Earth's surface rather than on the ground itself. The energy with which a near-Earth asteroid arrives at the top of the Earth's atmosphere is usually significantly more than the energy involved with its final air blast or impact. Much of the object's initial energy is lost to fragmentation or the heating of the atmosphere that surrounds its flight path toward the ground.

Land Impacts

If a near-Earth object is large enough to survive passage through the atmosphere intact and strike the Earth's surface, a number of destructive effects come into play. Near the blast site, or ground zero, there

---

[2] While the majority of meteorites are stony, small iron objects that are too strong to fragment can reach the ground with most of their energy remaining. However, these represent less than 5 percent of the mass that collides with Earth. Even so, iron meteorites are more easily recognized and more resistant to weathering so they represent a significant fraction of the meteorites currently in collections. Meteorites may be only a modest fraction of the small bodies that strike the atmosphere since most of them are probably too fragile to make it to the Earth's surface.

could be a blast wave, causing strong winds, a heat pulse, and seismic shaking. For the largest impactors, these local effects would pale in comparison to the global damage done by the fires started by the hot impact ejecta re-entering the Earth's atmosphere, the dust and soot thrown into the atmosphere, and acid rain. There would also be significant damage to the ozone layer and to the relative opacity of the atmosphere. The resulting loss of photosynthesis would produce an asteroid-induced winter.

Some of the energy of the impacting asteroid would be converted into a local blast wave and seismic shaking. However, most of this energy would be converted into heat or into the energy required to eject debris from the impact crater. The diameter of the impact crater would be about ten to fifteen times the diameter of the impactor itself, and impact shock waves would be sufficient to vaporize both the impactor and an amount of terrestrial material a few times the mass of the impactor. Far more material is pulverized by a land impact than is melted or vaporized.

An impactor larger than a few kilometers would cause a global catastrophe because of a brief initial heating of the Earth's atmosphere and surface followed by an extended period of cooling and relative darkness. Much of the impact ejecta would shoot back up the asteroid's incoming flight path in a ballistic fashion and exit the Earth's atmosphere along this path of least resistance. It would then reenter the atmosphere, heat to incandescence, and cause atmospheric heating and worldwide firestorms. These firestorms would then create enormous amounts of soot in the atmosphere. The combination of ejecta dust and firestorm soot could increase the opacity of the Earth's atmosphere for weeks at a time, shut down much of the sunlight and photosynthesis, and kill the plants as well as the creatures that depend upon these plants for food.

Ocean Impacts

Since more than two-thirds of the Earth's surface is covered with oceans, an impactor would most likely land in water. If the impactor diameter is larger than about 6 percent of the ocean depth, the ocean bottom would begin to crater. Water vapor, along with some crustal material, would be splashed into the atmosphere and beyond to space.

The greatest threat of a large ocean impactor may be the creation of an impact tsunami that could do significant damage to coastal regions far from the impact site. Ocean impacts have the potential to unleash tsunamis that could inundate coastlines. Since a large percentage of the population lives along the coasts, the damage to life and particularly that population's infrastructure could be a disproportionate contributor to the total near-Earth object hazard. The damage caused by an impact tsunami would depend upon the distance between the coast and the impact site, the depth of water at ground zero, and the nature of the coastal region. Waves lose their energy when they break, and they break when the wave height roughly equals the ocean depth. The most damage would be done if the wave did not break on a continental shelf before reaching the shoreline. If inhabited coastal regions were notified ahead of an arriving tsunami, an evacuation would be possible. Tsunami waves near shallow water coastlines do not move more than about twenty miles per hour so if he were warned, a beach bum on a bicycle might be able to outride it. However, coastal infrastructure damage could be significant.

Impactors larger than about one hundred meters that survive atmospheric entry in one piece might be expected to cause tsunamis. However, the generation of impact tsunamis is not yet well understood; only a modest amount of research has been carried out in this area. In addition, the geological record does not provide evidence for frequent tsunamis that are caused by impacts. Experiments were carried out in the 1960s to investigate whether nuclear weapons, when detonated in coastal waters, could be used as tactical weapons to generate tsunamis that would knock out enemy submarines along the coast of the United States. The results of these experiments showed that these types of impulse-induced waves broke at a considerable distance offshore on continental shelves and hence were ineffective as weapons. In addition, impact cratering in an ocean produces a much shorter wavelength feature that is likely to decay more rapidly than the long-wavelength tsunamis produced when vast tectonic plates move during earthquakes.

Regions without continental shelves, or with small ones, might still be at risk to impact-induced tsunamis, but a good deal more work needs to be done to investigate how the impact energy couples with ocean wave energy and how these impact tsunamis would propagate across the oceans and onto the shore.

Table 8.1 provides rough estimates of the population of stony near-Earth objects of various sizes. It also provides, for each object size, the typical energy with which the object would collide with the Earth's atmosphere measured in terms of tons, kilotons, or megatons of TNT explosives. The average interval between Earth impacts is given along with the expected final crater diameter if one of these objects survives passage through the Earth's atmosphere and suffers a ground impact into sedimentary rock.[3] The assumptions that were used to generate table 8.1 are 2.6 grams per cubic centimeter for the bulk density of the rocky asteroid impactors, 17 kilometers per second for an average impact velocity with the Earth's atmosphere, and 45 degrees as the most likely incoming flight angle with the local horizon. Stronger or denser asteroids or those on steeper trajectories will penetrate more deeply. Weaker and less dense objects or those coming in at more grazing angles will explode higher up.

During their trips through the Earth's atmosphere, most relatively small impactors lose much of their mass as they fragment and ablate in the atmosphere. For the same assumptions used to form table 8.1, the 1-, 10-, and 30-meter impactors would explode in an airburst at respective altitudes of about 50, 30, and 20 kilometers. The resistance of the atmosphere also slows them down to a terminal velocity. For example, although all of the impactors in table 8.1 were assumed to reach the Earth's atmosphere with a relative velocity of 17 kilometers per second, the respective terminal velocities before exploding for the 1-, 10-, 30-, and 100-meter-sized objects would be about 16, 13, 9, and 5 kilometers per second. The larger ones would impact the ground rather than explode in the atmosphere.

The largest impactor in table 8.1 has a diameter of ten kilometers, about the size of the object that caused the demise of the dinosaurs

---

[3] The estimates provided in table 8.1 were determined with the help of a spiffy website provided by Purdue University: http://www.purdue.edu/impactearth/.

TABLE 8.1. Average impact results, by size

Diameter of Impactor[1]	Total NEA Population[2]	Typical Impact Energy[3]	Average Interval between Impacts	Crater Diameter
1 m	1 billion	47 (8) tons	2 weeks	No crater
10 m	10 million	47 (19) KT	10 years	No crater
30 m	1.3 million	1.3 (0.9) MT	200 years	No crater
100 m	20,500–36,000	47 (4) MT	5,200 years	1.2 km
140 m	13,000–20,000	129 (49) MT	13,000 years	2.2 km
500 m	2,400–3,300	5,870 (5,610) MT	130,000 years	7.4 km
1 km	980–1,000	47,000 (46,300) MT	440,000 years	13.6 km
10 km	4	47 million MT	89 million years	104 km

[1] The smallest-diameter, stony, near-Earth asteroid that will reach the ground with at least half its original energy is about 160 meters. The diameter of the smallest rocky asteroid that can cause significant ground damage is in the range of 30 to 50 meters, roughly the size of the Tunguska impactor that fell in June 1908.

[2] The total near-Earth asteroid (NEA) population estimates, especially at the smaller sizes, are quite uncertain. For the objects whose diameters are about 100 meters, 140 meters, 500 meters, and 1 kilometer, the lower number was estimated using the NEOWISE infrared observations and the higher number was determined using the optical observations.

[3] For each size of impactor in the table, the first energy value given represents the total energy lost to the Earth's atmosphere and surrendered as a result of the final airburst or Earth impact. In each case, the energy value given in parentheses is the airburst or impact energy by itself. For example, for an impactor with an initial diameter of 140 meters, the total collision energy would be equivalent to 129 million tons (129 megatons) of TNT explosives, but 80 megatons of this energy would be lost due to the object fragmenting and heating the atmosphere during its flight path. That would leave 49 megatons for the final Earth impact itself.

sixty-five million years ago. However, currently the biggest potentially hazardous asteroid that can approach the Earth's orbit is (4179) Toutatis, whose largest dimension is 4.6 kilometers. Either the dinosaur-killing asteroid was unusually large or perhaps the culprit was a large comet arriving from the distant Oort cloud.

Near the small end of the impactors noted in table 8.1 are the basketball- and larger-sized Earth impactors that occur daily, causing fireball events that can be spectacular celestial displays. But most of these events go unnoticed, occurring as they do over the oceans, over uninhabited regions, or when most of us are sleeping. However, they do not escape the notice of down-looking Department of Defense infrared and visible light sensors in orbit about the Earth. These eyes in the sky, designed to detect missile launches and nuclear explosions, observe at least one fireball event every few days. Volkswagen-sized impactors occur a few times each year and can cause quite a stir as they fragment and descend into the Earth's atmosphere. Some significant fireball and impact events include the Grand Teton fireball of August 10, 1972, the Great Fireball of February 1, 1994, the Carancas impactor of September 15, 2007, the Tunguska air blast event of June 30, 1908, and the Chesapeake Bay impact event that occurred thirty-five million years ago.

Significant Fireball and Impact Events

The Grand Teton Fireball Event of August 10, 1972

A small fireball object, perhaps three meters in diameter, was widely observed during daylight on August 10, 1972, when it flew south to north from Utah through Idaho, Montana, and into Canada. It is sometimes referred to as the Grand Teton fireball event for the number of people who witnessed it over this impressive Wyoming mountain range. Entering the Earth's atmosphere at a fairly low angle with respect to the horizon and moving too fast to be captured by the Earth, this near-Earth asteroid penetrated down to about sixty kilometers altitude before flying back into space. It lost some of its mass and was slowed down a bit but otherwise suffered little from its close-Earth encounter.

The Great Fireball of February 1, 1994

The largest fireball event recorded by satellite to date occurred on February 1, 1994, over the South Pacific. Had you been on the deck of a cruise ship at the time, you would have noted the fireball descending into the Earth's lower atmosphere with a brightness rivaling that of the noonday Sun. The event's total energy, as measured by Defense Department infrared sensors, was about sixty kilotons of energy; using table 8.1, we can infer that the original size of the impactor was probably just over ten meters in diameter. One could expect an event of this magnitude every ten years on average.

FIGURE 8.1. The Carancas, Peru, impact crater, about fourteen meters in diameter, was apparently caused by the impact of a small one-meter-sized asteroid on September 15, 2007.
Source: Courtesy of Peter H. Schultz, Brown University.

The Carancas Event of September 15, 2007

A small asteroid is given credit for creating nearly a fourteen-meter diameter crater near the village of Carancas, Peru, near Lake Titicaca and the Bolivian border. The impact took place at 11:45 AM local time. The explosion was heard twenty kilometers from the impact site and caused broken window glass in a local health center one kilometer from ground zero. It was reported that soon after the impact some of the villagers became ill for a few days, perhaps due to the toxic nature of the sulfurous groundwater that quickly seeped into the crater and began to bubble as a result of the impact energy. A one-meter-sized meteorite likely caused this fall, but there were no reports of it being a breakup fragment from a larger parent body. Small five-centimeter ordinary chondrite meteorite fragments were found two hundred meters from the crater and if, indeed, this small object survived the atmosphere without fragmenting, it was unusually tough or could have been long and skinny and entered the atmosphere pointy-end first.

The Tunguska Event

On June 30, 1908, at 7:17 AM local time, an extraordinary air blast explosion took place above the Tunguska region of Russian Siberia.

FIGURE 8.2. During the early morning hours of June 30, 1908, in Russian Siberia, an asteroid collided with the Earth's atmosphere causing an air blast event that leveled millions of trees over an area that spanned 2,200 square kilometers. The area was very sparsely populated and no injuries to humans were reported.

Trees were burned out to a few kilometers from ground zero and knocked down in a butterfly-shaped pattern over an area spanning 2,200 square kilometers. Eyewitness accounts disagree on both the time and the direction of the flight of the near-Earth object, but geomagnetic disturbances were noted at Irkutsk, 900 kilometers southeast of the blast, and Tashkent, Tbilisi, and Jena. A seismograph station in St. Petersburg, 4,000 kilometers away, noted a disturbance, as did more distant stations around the world. The explosion registered on seismic stations across Eurasia, reaching 5.0 on the Richter scale. Bright nights caused by a great deal of dust in the atmosphere reflecting sunlight to nighttime regions occurred in northern Europe on the night of the Tunguska event and for a few nights following. During these nights the sky glowed so brightly that it was possible to read a newspaper at midnight. Although there was extensive ground damage, no obvious crater was located below the blast site. The region is so remote that an investigative team, led by the Russian scientist L. A. Kulik, did not reach the site until the spring of 1927.[4]

The most likely cause of the Tunguska event was the impact of a small asteroid about forty meters in diameter that created an above ground airburst releasing three to five megatons of equivalent energy. Until recent work by Mark Boslough and his colleagues, the energy of the blast had been thought to be ten to fifteen megatons of energy based upon above ground nuclear tests and the resulting ground damage. However, Boslough's detailed computer simulations suggest that this was not a point source explosion at a specific altitude but an explosion that was hurtling downward at high velocity and hence brought more of the blast's momentum to the Earth's surface.

Forty-meter-sized asteroids would be expected to strike the Earth every few hundred years; thus an event like this more than one hundred years ago seems plausible. However, there have been a plethora of alternate suggestions for the Tunguska event. These range from the unlikely to the ridiculous and include a colliding comet, mini black hole, antimatter, the crash of an alien spacecraft, and an overzealous laser signal from an advanced race holed up near the star 61 Cygni. The local inhabitants of this very sparsely populated region believed that the blast was a visitation by the god Ogdy, who had cursed the area by smashing trees and killing animals.

The Chesapeake Bay Impact Event

Inhabitants of the Chesapeake Bay area of Virginia and Maryland refer to this region as the land of pleasant living, but there was nothing pleasant there about thirty-five million years ago when an asteroid three to five kilometers in diameter smashed into the area near what is

---

[4] In 1920, L. A. Kulik was put in charge of the meteorite collection at the St. Petersburg museum and was tasked with locating and examining the meteorites that had fallen in the Soviet Union. He made a first attempt to visit the Tunguska site in 1921 but was unsuccessful. He managed to make on-site investigations in 1927, 1929, and 1938 when he photographed and documented the area. He spent some time in prison for revolutionary activities and died in a Nazi prison camp in April 1942.

now Cape Charles, Virginia, near the southern end of the Delmarva peninsula. The initial impact coupled with a train of tsunami waves created a crater about eighty-five kilometers in diameter. Historic Williamsburg and Jamestown are located on the outer edges of this impact zone. The evidence for the impact is convincing. Cores drilled into the region reveal the shocked quartz that forms as a result of such energetic impacts and the fossil layers are jumbled with older specimens above younger ones. The ancient layers of sand and water, called aquifers, that would normally provide the region's water supply cannot always be relied upon for fresh well water since many of them have been disturbed and are now salty.

The Stealth Asteroids and Show-Off Comets

While cometary apparitions have been recorded for millennia, it wasn't until 1801 that Ceres, the first asteroid, was discovered, 1898 when Eros, the first near-Earth asteroid, was discovered, and 1932 when Apollo, the first potentially hazardous asteroid, was discovered. Comets are show-offs, belching easily seen gas and dust as they near the Sun while the far more stealthy asteroids are relatively dark and rarely observable with the naked eye.[5]

Long-period comets, defined here as active comets with orbital periods greater than two hundred years, are the most difficult objects to mitigate should one be found on an Earth-threatening trajectory. The arrival of these objects from the outer solar system cannot be predicted and the impact warning time would be measured in a few months, not years. In general long-period comets do not become active, and hence discoverable, until inside the orbit of Jupiter, and it takes but nine months for them to travel the distance from Jupiter's orbit to that of

---

[5] Naked-eye visibility is usually defined as magnitude 6 or brighter (i.e., magnitudes less than 6). Vesta occasionally reaches magnitude 5.3 and can be seen in a clear, dark sky by someone with good eyesight if he knows where to look. The next time that Vesta will reach magnitude 5.3 is on July 10, 2029. Three months earlier, on April 13, 2029, near-Earth asteroid Apophis will pass within 5 Earth radii of the Earth's surface and be observable with the naked eye at apparent magnitude 3.5 (5 times brighter than Vesta) in Europe and North Africa. Mark your calendar.

the Earth. An impact energy is proportional to the object's mass, and hence its bulk density, as well as the square of its impact velocity. A typical impact velocity of a long-period comet would be about 51 kilometers per second, three times the 17 kilometers per second value for a typical near-Earth asteroid. Hence the impact energy for a long-period comet would be nine times that of a near-Earth asteroid of similar mass. However, the bulk density of a comet, which is about 0.6 grams per cubic centimeter, is several times less than the density of a stony asteroid, which is about 2.6 grams per cubic centimeter, so for a long-period comet and a stony near-Earth asteroid of the same size, the comet's impact energy would be about twice that of the asteroid.[6]

The Earth-impact rate for active comets is less than 1 percent the near-Earth asteroid impact rate. There are a few lines of evidence that support this claim. Zdenek Sekanina and I looked at all the recorded comets between 1300 and 2000 that had reliable orbits. We noted the number that made close-Earth approaches to within certain distances from the Earth. From our analysis, we were able to deduce that a long-period comet would be expected to strike the Earth only once every forty-three million years on average, far less frequent than asteroid strikes. Alan Chamberlin and I carried out another more direct technique to determine the relative number of comets versus asteroids in the near-Earth object population. We noted that during the interval from 1900 through January 2011, 2,460 known near-Earth asteroids of all sizes made 3,901 close-Earth approaches to within 0.05 AU. During the same interval, only three known Jupiter family comets (7P/Pons-Winnecke in June 1927, 1999 R1 SOHO in April 1947, and 1999 J6 SOHO in June 1999) and one long-period comet (1983 H1 IRAS-Araki-Alcock in May 1983) have come as close. No Halley-type comets came this close. Hence, when compared to near-Earth

---

[6] Very few asteroids have had their masses and volumes measured, so their bulk densities (mass divided by volume) are unknown. Using spacecraft data, the bulk density of stony, near-Earth asteroid Eros was determined to be 2.7 grams per cubic centimeter. The first accurate measurement of a comet's bulk density will have to await the arrival of the European Space Agency's Rosetta spacecraft at comet 67P/Churyumov-Gerasimenko in 2014, but indirect measurements of comets suggest their bulk densities are in the neighborhood of 0.6 grams per cubic centimeter. Since water has a bulk density of one gram per cubic centimeter, if we had a bowl of water large enough to hold a cometary nucleus, it would float.

asteroids, the number of Earth impacts by all active comets, including Jupiter family comets, Halley-type comets, and long-period comets, is well below 1 percent. However, at the upper end of the impactor size ranges, the number of large comets capable of impacting Earth is comparable to the number of large near-Earth asteroids capable of doing the same.

Asteroid Threat? What Asteroid Threat?

In a 2003 NASA study it was concluded that the consequences of a major impact event to civilization would be so costly that the ongoing ground-based surveys should be continued. The assessed risk of a particular object is reduced if it is discovered and tracked for years and found to be harmless. In the very unlikely event that a particular large object were to be found on an Earth-threatening trajectory, there would likely be time to mount a spacecraft mission to deflect it. In either case, the assessed risk of this object would be dramatically reduced for decades. In this regard, the ongoing NASA Near-Earth Object Observations Program to discover and track these objects has been an outstanding success.

The success of the near-Earth object survey programs to date raises the question of what more needs to be done. The current NASA goal is to find and track 90 percent of the potentially hazardous near-Earth objects larger than 140 meters in diameter and physically characterize a representative sample. Less than half of this population has been discovered to date. Using current telescope assets, it could take many years to meet this goal. How, then, does the current risk from near-Earth objects compare to other risks? The largest undiscovered objects always dominate the risk from near-Earth asteroids. An impact by an object a kilometer or two in diameter could kill about a billion people and happen once in a million years on average. Hence the average number of fatalities per year over very long time intervals would be about one thousand. The current near-Earth object search efforts have already found more than 90 percent of this population and none of them is a current threat for at least one hundred years. Thus the leftover short-term risk is roughly one hundred fatalities per year.

TABLE 8.2. The average number of fatalities per year worldwide from various accidents, events, and illnesses, compared to the estimated 100 annual fatalities for near-Earth object collisions

Threat	Estimated Fatalities per Year
Shark attacks	3–7
Asteroids	91
Earthquakes	36,000
Malaria	1 million
Traffic accidents	1.2 million
Air pollution	2 million
HIV/AIDS	2.1 million
Tobacco	5 million

Note: The estimate of 91 annual deaths is worse than that for shark attacks and about comparable to that of accidents due to fireworks. On the other hand, fireworks do not have the capacity to reduce large regions of the Earth's surface to ashes and create another extinction event. Comparisons like those in this table are a bit misleading because near-Earth asteroid impacts are very low-probability—but very high-consequence—events. Obviously 91 people do not die each year from asteroid impacts; this is a long-term average arising from very catastrophic events occurring only rarely.

This figure will drop as more and more near-Earth objects are discovered.

Much of the data in table 8.2 are presented in a 2010 National Research Council report. Included within that report is a minority opinion by Mark Boslough, a scientist from Sandia Laboratory in Albuquerque, New Mexico. Boslough argues that any table akin to table 8.2 should include an entry of 150,000 fatalities per year due to long-term climate change. This estimate comes from the World Health Organization, and Boslough noted that estimates of fatalities from a catastrophic near-Earth object impact are largely derived from computer models developed to predict the effects of climate change. However, Boslough's arguments were not included in the main body of the report since it was thought that reliable estimates were not available and this topic would distract from the issue at hand.[7]

---

[7] Mark Boslough's well-respected scientific chops are matched by his keen sense of humor that he occasionally employs to skewer pseudoscience. In a 1998 April Fool's Day spoof on creationists, he wrote about a supposed vote in the Alabama legislature to change the value of pi to its "biblical value" of exactly three. For a time, this story was widely believed.

Things That Go Bump in the Night

There are no recorded and verified human fatalities directly attributable to near-Earth objects to date. However, automobiles have been destroyed, buildings have suffered, and a young boy from Uganda was bonked harmlessly by a tiny meteorite fragment in August 1992. Ancient Chinese records do report cases where falling rocks and irons caused fatalities and the destruction of dwellings, but it is difficult to judge the veracity of these accounts.

One of the few (perhaps the only) documented cases of a person getting hit by a meteorite in North America took place on November 30, 1954, when a stony meteorite weighing in at 3.9 kilograms penetrated the roof of a house in Sylacauga, Alabama, bounced off a radio, and struck a woman who was asleep on a nearby couch. She suffered painful bruises on her left hip and arm.

By estimating the number of meteorites that do impact the Earth's surface, we can also estimate the likelihood of a meteorite striking a human. In North America, the estimate is an average of once every 180 years (and about once per year for an impact upon a building). For the entire Earth's surface, one would expect one human impact every nine years on average with about sixteen buildings per year taking a hit. Because there are so many more of them, small asteroids are far more likely to strike Earth than large ones. However, it is important to keep in mind that while Earth impacts by large near-Earth objects are rare with very low odds of hitting anything, such events do happen and would have catastrophic consequences. We cannot afford to play the odds when civilization is at stake.

✧ CHAPTER 9 ✧

Predicting the Likelihood of an Earth Impact

It's tough to make predictions, especially about the future.
—attributed to Yogi Berra

"Hello, White House? There's an Asteroid Headed toward Earth"

During the morning of October 6, 2008, Eastern Standard Time (EST), Tim Spahr, director of the Minor Planet Center (MPC), couldn't believe what his computer was telling him. In less than twelve hours, a near-Earth asteroid would collide with Earth. Spahr had just received observations of a fast-moving, near-Earth asteroid discovered by Richard Kowalski at the Catalina Sky Survey near Tucson, and his preliminary trajectory computations, based upon these observations, suggested a nearly certain and imminent encounter with Earth. He quickly notified both Lindley Johnson at NASA headquarters and Steve Chesley at the Jet Propulsion Laboratory (JPL) and posted the orbit on the MPC's Near-Earth Object Confirmation Page on the Web. Twenty-six amateur and professional astronomers, who keep an eye on this page, quickly pointed their telescopes toward the incoming asteroid, now designated 2008 TC3, and began reporting additional observations to the MPC. These invaluable, additional observations were then forwarded to JPL, where Steve Chesley refined the object's orbit and provided a series of improved predictions for the time and the location near the Earth's surface where the impact would occur. An airburst somewhere over northern Africa was predicted. From the object's faintness and proximity to Earth, it was clear that it was only a few meters across and would likely not cause ground damage. However, the precise time and location of the airburst were important to quiet fears near the affected location and alert scientists of a possible shower of precious meteorites that could be used to study the chemical composition of the parent asteroid.

Within an hour of receiving the initial data set, JPL predicted that the object would enter the Earth's atmosphere above northern Sudan around 02:46 Greenwich time on October 7 (5:46 AM local time in Sudan). As the day progressed and more and more observations arrived at the MPC and were forwarded to JPL, Steve Chesley and Paul Chodas continued to improve the orbit for 2008 TC3 and forwarded updated predictions to the astronomical community and to NASA headquarters. Though the estimated size would only produce a fireball and no damage at ground level, just to be safe, NASA headquarters alerted officials at the National Security Council, the Office of Science and Technology Policy, the Department of State, the Department of Homeland Security, the National Military Command Center at the Pentagon, and the Air Force Space Command's Joint Space Operations Center. NASA then issued a press release announcing the predicted impact to the public. An e-mail with the NASA announcement of the coming impact even made it to the White House Press Office where it was noted by Dana Perino as the most unusual e-mail she had received while serving as press secretary for President George W. Bush.[1]

A spectacular fireball did light up the predawn sky above northern Sudan at exactly the predicted time and was witnessed by a KLM airline pilot while flying over Chad. The explosion scattered small meteorite fragments across the Nubian Desert below. Although such small-impact events occur several times per year around the globe, this

---

[1] About 9:30 EST on October 6, 2008, Ms. Perino received an e-mail with a subject line of "HEADS UP." It warned that an asteroid was headed toward Sudan and suggested that the Sudanese be contacted to let them know it was coming, but no such call was made because there were no formal relations between the United States and the Sudanese government.

case was unprecedented because the asteroid was actually discovered before it reached the Earth and the impact location and time were, for the first time, predicted in advance.

Detections of the actual atmospheric impact event suggested that it was an airburst explosion at an altitude of thirty-seven kilometers with an energy equivalent of about one kiloton of TNT explosives. The predicted impact time and place agreed very well with a number of atmospheric entry observations, including those from U.S. government missile warning systems, low-frequency infrasound signals from two ground stations, and images from the Meteosat 8 weather satellite. The latest JPL trajectory estimate, which carefully considers all available data, including some measurements not available until after the event, was accurate to within a few kilometers at the time of atmospheric entry.

This dramatic prediction of an actual impact underscored the successful evolution of the Near-Earth Object (NEO) Program's discovery and orbit prediction process. The discovery of a four-meter-sized asteroid was made while it was located at 1.3 lunar distances. Twenty-six international observatories provided data within hours while the object was still inbound, and the orbit and impact computations were determined, verified, and announced well before the impact, which took place only twenty hours after the discovery itself. The altitude of the air blast was at thirty-seven kilometers and the predicted time and location of the air blast agreed with the observations to within one second and to within one-tenth of a degree in longitude and latitude. The NEO impact warning system worked very well for this, the first predicted impact by an NEO.

Using the ground track positions provided by Steve Chesley and Paul Chodas of JPL, a few months later meteorite recovery expeditions led by Peter Jenniskens of the SETI Institute in California and Muawia H. Shaddad of the University of Khartoum successfully retrieved hundreds of meteorites totaling about four kilograms from this near-Earth asteroid. Although these fragments were mostly composed of partially melted carbonaceous chondritic material classified as relatively rare Ureilite meteorites, they had materials more representative of other more common types of meteorites as well. This was an unexpected

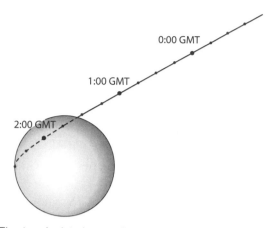

Fɪɢᴜʀᴇ 9.1. The terminal trajectory for Earth-impacting asteroid 2008 TC3. The view is from the Sun. Note that the asteroid entered Earth's shadow at about 1:49 AM Greenwich time on October 7, 2008 (8:49 PM EST on October 6), so that the final portion of the trajectory is in shadow behind the Earth.

result since a small asteroid would be expected to be rather homogeneous in its composition. Apparently there was some mixing of composite materials, perhaps caused by long-ago impacts by other more common types of asteroids and a subsequent reassembly into a second-generation asteroid. These asteroid fragments are now referred to as the Almahata Sitta meteorites. Almahata Sitta is Arabic for Station Six, an isolated railway station, truck stop, and teahouse located near the meteorite recovery site in northern Sudan.

The Orbit Determination Process

The orbit determination process begins with only a few observations, usually those used to discover the object. The Gauss technique, used so successfully during the discovery of the first asteroid, Ceres, in 1801, is an example of a preliminary orbit determination method. Once a preliminary orbit is determined, subsequent observations are used to update, or improve, this orbit by adjusting the initial orbit parameters until the object's computed trajectory can successfully predict the positions where the object was actually observed. That is, the predicted positions of the object computed using the initial orbit are compared to the actual celestial positions observed at specific times. The initial orbit is then incrementally adjusted until the differences, or residuals, between the observed positions of the object at the observation times and the predicted positions of the object at those same times are minimized. Once the initial orbit has been established based upon optical observations of the object's position on the night sky, radar observations are made when possible and used to dramatically improve the object's orbit. The initial orbit includes all six orbital elements that are valid for a particular instant of time, or epoch. Given these starting conditions, the object's trajectory can be carried forward or backward in time on computers by taking into account the subtle gravitational nudges of neighboring planets, some of the larger asteroids, the Moon, and the very subtle effects of cometary outgassing or the Yarkovsky thermal re-radiation effects acting upon small asteroids.

An object's calculated orbit gives only its most likely, or nominal, position at any given time so its possible positions at any given time are represented by an uncertainty ellipsoid that surrounds the nominal position. Think of the object's nominal, or most likely, position as the center of a football with any number of less likely but acceptable alternative positions occupying other locations within the football-shaped uncertainty ellipsoid. As one moves away from the central or nominal position and gets to either end of the football, the likelihood of the object occupying that position becomes less and less. As we extrapolate the object's motion further into the future, its position uncertainties usually increase with time. Imagine that in its motion about the Sun, the object's uncertainty ellipsoid, or football, gets more and more elongated with time. If any part of that elongated ellipsoid should touch the Earth at some time in the future, an Earth impact cannot be ruled out. There is then a non-zero impact probability.[2] If we then imagine a

---

[2] Strictly speaking, there is a non-zero impact probability if any portion of the uncertainty ellipsoid touches the Earth's capture cross section, which is somewhat larger than the Earth's diameter and depends upon the relative velocity between the NEO and Earth. For example, if a NEO approaches the Earth with a relative velocity of 10 km/s, it only needs to be aimed within 1.5 Earth radius of the Earth's center for the Earth's gravitational attraction to cause a collision.

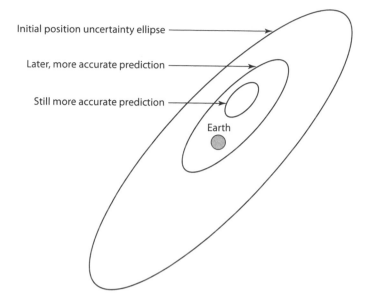

Initial position uncertainty ellipse

Later, more accurate prediction

Still more accurate prediction

Earth

FIGURE 9.2. The Earth's target plane diagram. At the time of a close-Earth approach, an object's position uncertainty ellipsoid, when projected on a plane perpendicular to the incoming trajectory, becomes an uncertainty ellipse, and if any portion of this ellipse should touch or include the Earth's capture cross section, there will be a chance of an Earth impact. An object's uncertain initial orbit will provide uncertain future predictions for its position during a close-Earth approach. Then its relatively large initial uncertainty ellipse could include the Earth and an impact could not be ruled out. However, the object's future position predictions improve as more observations are included in the orbit updates: the object's uncertainty ellipse at the time of the close-Earth approach would then shrink and, almost always, subsequent orbit updates would completely rule out an Earth collision.

plane drawn through the Earth and perpendicular to the object's flight path with respect to the Earth, the three-dimensional uncertainty ellipsoid will project onto the Earth's impact plane as a two-dimensional uncertainty ellipse. The portion of the uncertainty ellipse that touches the Earth divided by the total area of the uncertainty ellipse is the impact probability. If the entire uncertainty ellipse of the object projects wholly onto the Earth, the impact probability would then be 1 or 100 percent. Lights out.

However, as more and more optical and radar observations are taken, the orbit of a near-Earth object can be refined with the result that the uncertainty region surrounding the nominal position of the near-Earth object is gradually reduced. Then this region will probably not touch the Earth in the future. Thus the non-zero impact probabilities that can arise for recently discovered near-Earth objects with uncertain orbits will generally become less and drop to a 0 percent chance of an impact as more and more observations are included in the orbit determination process. The lesson here is that we need to continually re-observe and track the subset of near-Earth objects that are potentially hazardous.

NASA's Near-Earth Object Program Office

In July 1998 NASA established the Near-Earth Object (NEO) Program Office at JPL to coordinate and monitor the discovery of NEOs and their future motions and to compute close-Earth approaches and, if appropriate, their Earth-impact probabilities. In March of the following year, JPL's NEO Program Office first posted its website, providing information on near-Earth objects including orbital data, ephemeris information, and upcoming close approaches, as well as any known physical data.[3]

JPL's NEO Program Office receives astrometric data and preliminary orbits from the Minor Planet Center (MPC) and then continuously improves these orbits, and the resulting close-Earth approach predictions, as additional data are received. Once a new orbit has been successfully fitted to the available observational data, the object's trajectory is numerically integrated forward in time to note any close-Earth approaches in the next hundred years. The JPL orbital computations employ state-of-the-art numerical computer models that take into account the gravitational perturbations by the planets, the Moon, and large asteroids, as well as relativistic and thermal re-radiation or outgassing effects. These updated orbits and close-approach information are

---

[3] The website address for JPL's NEO Program Office is http://neo.jpl.nasa.gov. A less technical site for near-Earth objects is http://www.jpl.nasa.gov/asteroidwatch/index.cfm.

automatically computed and immediately posted to the NEO Program Office website. Those objects for which an Earth impact cannot yet be ruled out are automatically submitted to the Sentry system for further risk analysis.

Within Sentry, the possible future orbits of an object are examined and Earth-impact probabilities computed for specific future dates. These results are immediately posted to the JPL NEO website. The only exception to this chain of events occurs if relatively large objects that have relatively high impact probabilities and short time intervals to possible Earth impacts are discovered by the Sentry system. In this latter scenario, an e-mail message is sent to the NEO Program Office personnel requesting verification of the events before the information is posted to the website. This manual verification process then includes electronic correspondence with colleagues at Pisa to compare results and, if verified, notification of these results to NASA headquarters. Additional verification at JPL is also conducted by an independent process that determines thousands of slightly different variant orbits that could be used to successfully fit the available observations and then numerically integrates each orbit forward to the time of the possible Earth impact. In this so-called Monte Carlo process, the number of these variant orbits that impact Earth, compared to the total number of variant orbits, gives a rigorous Earth-impact probability. This computationally intensive Monte Carlo process is used to verify the results from the faster Sentry system.

Occasionally small near-Earth objects are reported as new discoveries when in actuality they are man-made objects. The S-IVB rocket booster stages of Apollo 8, 9, 10, 11, and 12 all wound up in heliocentric orbits interior to the Earth's orbit. There are any number of small near-Earth asteroids in very similar orbits; Paul Chodas has been called upon more than once to investigate whether the trajectory of a recently found near-Earth object has made previous close-Earth approaches around the time of Apollo launches to discern whether the new object is really a natural body or possibly an Apollo rocket stage. A small near-Earth object discovered by the astronomer Bill Yeung on September 3, 2002, was found to be on a temporary Earth orbit. Chodas investigated its motion and suggested that it returned to a heliocentric orbit in June

2003 and was most likely the S-IVB stage of the Apollo 12 spacecraft. When spectral measurements of the object were taken, they were consistent with titanium dioxide paint, the same paint used by NASA for the Saturn rockets.

Predicting Earth Encounters over Long Time Intervals

The Sentry system at JPL and the NEODyS system in Pisa compute near-Earth object trajectories out to about one hundred years from the current date. In general, going out further in time would introduce substantial uncertainties in the object's predicted trajectory; this is especially true if the object should pass closely by a planetary body in the future. Often the uncertain encounter conditions during one of these close planetary approaches introduce very substantial orbital uncertainties, making subsequent accurate predictions difficult. However, for a few asteroids with very good orbits, well refined by long optical observational intervals and perhaps with powerful radar data, orbital extrapolations can be carried out well beyond one hundred years. Near-Earth asteroids 1950 DA and 1999 RQ36 are two such cases.

Near-Earth asteroid (29075) 1950 DA is thought to be about one kilometer in diameter. Its excellent orbit, with an uncertainty currently in its semi-major axis of only about 100 meters, is due to extensive optical observations that have been made since 1950 along with radar observations that were carried out in March 2001. In a 2002 study led by Jon Giorgini, a remote Earth-impact possibility was identified for March 16, 2880, nearly nine centuries from now. Because of the long time interval between now and 2880, Giorgini and his colleagues examined several subtle perturbative effects that are not normally included in projecting the motions of asteroids forward in time. These effects include the gravitational tugs of the stars in the Milky Way galaxy, several asteroids, the oblate Sun whose mass is decreasing with time, the uncertainties in the planetary masses, and the uncertainties in the computational process itself. They also investigated the pressure on the asteroid induced by the solar wind, a stream of charged particles that emanates from the Sun, the pressure of sunlight, and the thermal re-emission of sunlight known as the Yarkovsky effect. This study concluded that the Earth-impact probability in 2880 depends largely upon the unknown Yarkovsky effect, which in turn depends upon the asteroid's rotation characteristics and its unknown physical properties. For example, if the asteroid is rotating in the same direct sense as its motion about the Sun, then the Yarkovsky effect will add orbital energy to the asteroid, causing its orbital period to increase a bit and moving it and its associated uncertain region closer to the Earth in 2880. In this case, the probability of an Earth impact could be as high as one in three hundred if the asteroid's reflectivity, mass, and surface properties are just right. However, by far the most likely scenario has the asteroid missing the Earth by a wide margin in March 2880.

A multiyear optical observation interval and radar observations on three separate occasions in 1999, 2005, and 2011 have shown that near-Earth asteroid (101955) 1999 RQ36 also has a very secure orbit, with an uncertainty currently in its orbital semi-major axis of only several meters. Radar observations have been used to determine that the object is about 500 meters in diameter and rotates once every 4.3 hours in a retrograde fashion. According to studies by Andrea Milani, Steve Chesley, and their colleagues in 2009 and 2011, there are a few remote Earth-impact possibilities in the late twenty-second century. Once again, the most likely scenario by far is that with additional observations to refine the orbit, this asteroid will be shown to miss the Earth by comfortable margins in the late twenty-second century. But it is an object to keep an eye on, and NASA is doing just that.

In May 2011, NASA selected for flight the OSIRIS-REx mission to rendezvous with near-Earth asteroid 1999 RQ36 in 2020 and return to Earth a surface sample in 2023. Using its onboard suite of instruments that includes three cameras and three spectrometers, the spacecraft should provide a detailed description of this dark carbonaceous asteroid. When the returned samples are intensively studied with ground-based instrumentation, the results should shed light on the extent to which these types of near-Earth asteroids delivered to the early Earth the carbon-based materials, or organics, that allowed life to form.[4]

---

[4] OSIRIS-REx is a contrived acronym for the rather convoluted name "Origins-Spectral Interpretation-Resource Identification-Security-Regolith Explorer." Until his untimely death

Apophis: The Poster Child of Near-Earth Objects

On Friday the thirteenth of April 2029, a near-Earth asteroid with a diameter the size of the entire Rose Bowl football stadium will pass within 5 Earth radii of the Earth's surface, briefly appear as a naked-eye object, and dramatically focus the world's attention on a shot across the bow by Mother Nature. Asteroid (99942) Apophis will actually pass within the distance of the communications satellites that will be busy broadcasting its arrival.

Apophis was discovered on June 19, 2004, by Roy Tucker, David Tholen, and Fabrizio Bernardi of the NASA-funded University of Hawaii Asteroid Survey from Kitt Peak, Arizona, and observed over two nights. This was not enough of an observation span to secure its orbit so it was then lost. However, on December 18, Gordon Garradd of the Siding Spring Survey, another NASA-funded effort, rediscovered the object from Australia. Further observations from around the globe over the next several days allowed the Minor Planet Center to link the December observations with the earlier observations from June, thus identifying them as pertaining to a unique asteroid. It was then that the possibility of impact in 2029 was realized by the automatic Sentry system of NASA's Near-Earth Object Program Office. NEODyS also detected the impact possibility and provided similar predictions.

For a few days around Christmas 2004, the impact probability calculations for April 13, 2029, a Friday, reached as high as one chance in thirty-seven, but the Christmas Day angst was removed two days later when heretofore unrecognized pre-discovery observations of this object were located in the Spacewatch data archives by Jeff Larsen and Anne Descour. When included in an updated orbit, these observations greatly narrowed the uncertainties and completely ruled out the possibility of an Earth impact in 2029. Arecibo radar observations taken in late January 2005 allowed a further refinement of the orbit of Apophis, and while the new orbit actually showed Apophis would pass closer than 6 Earth radii from the Earth's surface, there was still no chance for

---

in September 2011, Michael Drake from the University of Arizona was the Principal Investigator for this mission. The mission management is being carried out by the Goddard Space Flight Center.

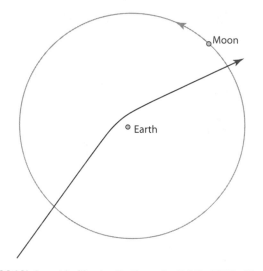

F<small>IGURE</small> 9.3. (99942) Apophis flies by Earth on April 13, 2029. The most likely trajectory of Apophis is shown as a curved line that passes near the Earth on April 13, 2029. The collection of possible Apophis orbits that make up its position uncertainty ellipsoid do not reach the Earth's capture cross section so no Earth impact is possible in 2029.

an Earth or Moon impact in April 2029. The passage of the asteroid by the Earth in 2029 alters its subsequent trajectory and causes its position uncertainty region to expand rapidly as it moves away from Earth. As a result, the asteroid's motion is much less predictable after the 2029 close-Earth approach and, initially, future Earth impacts were still remotely possible in 2034, 2035, and 2036. Subsequent orbital refinements based upon additional observations have ruled out the 2034 and 2035 possibilities.

While the Earth impacts in 2029, 2034, and 2035 are no longer possible, there is still a remote chance that when Apophis closely passes by the Earth in 2029, its trajectory will be modified just enough to cause it to return after six more orbits around the Sun and collide with the Earth on April 13, 2036, which will be Easter Sunday. This potential collision in 2036 will only occur if Apophis, during its 2029 close-Earth flyby, passes through a 610-meter-sized region in space that allows the gravitational effect of the nearby Earth to modify the orbit of Apophis just enough and in the right direction to set up an impact exactly seven years later. This predicted tiny region in space, termed a "keyhole" by Paul Chodas, provides interesting leverage in the unlikely event that the 2036 Earth impact becomes a real possibility. Any spacecraft deflection attempt completed prior to the 2029 close-Earth approach would not have to move Apophis by even an Earth radius (6,378 kilometers) to avoid the Earth impact in 2036. Rather the deflection attempt need only move Apophis less than one ten-thousandth of this distance—just out of the 610-meter-sized keyhole during the 2029 close-Earth approach. As is the case for almost all potential impacts identified by the Sentry and NEODyS systems, additional observations of Apophis will be used to refine its orbit, reduce its uncertainty region, and almost certainly rule out any Earth impact in 2036.

The Cosmic Sucker Punch: Unexpected Impacts by Undiscovered Objects

Asteroid 2011 CQ1 was discovered by the Catalina Sky Survey on February 4, 2011, and made a near record close-Earth approach fourteen hours later on February 4 at 19:39 UT or 14:39 EST. It passed to within 0.85 Earth radii, or 5,480 kilometers, of the Earth's surface over a region in the mid-Pacific. This object, only about one meter in diameter, is one of the closest non-impacting objects in our asteroid catalog to date. Prior to the close-Earth approach, this object was in a so-called Apollo-class orbit that was mostly outside the Earth's orbit. Following the close approach, the Earth's gravitational attraction modified the object's orbit to an Aten-class orbit, so now the asteroid spends almost all of its time inside the Earth's orbit.

The close-Earth approach changed the asteroid's flight path by 68 degrees. Because of their diminutive sizes, objects like this are extremely difficult to discover, but there is likely to be a billion of them this size and larger in near-Earth space; we could expect one to strike Earth's atmosphere every few weeks on average. Upon striking the atmosphere, small objects of this size create visually impressive fireball events, but only rarely do even a few small fragments reach the ground.

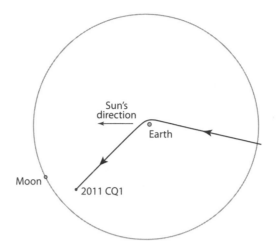

Figure 9.4. Near-Earth asteroid 2011 CQ1, a one-meter-sized asteroid, flew within 0.85 Earth radii of the Earth's surface on February 4, 2011. The Earth's gravitational attraction bent its orbit by 68 degrees before the asteroid resumed its journey about the Sun.

Most near-Earth objects smaller than about thirty meters in diameter would not be expected to cause significant ground damage. However, there are more than a million near-Earth objects larger than thirty meters and we've discovered less than 1 percent of this population. NASA's focus to date has been on finding the much larger objects that can cause widespread, even global disasters. This is as it should be but the fact remains that the vast majority of near-Earth asteroids that could cause ground damage remain undiscovered. And, as we have seen, predicting the impacts of the even more numerous near-Earth objects down to a few meters in size, like 2008 TC3, could produce additional scientific bonanzas of meteorites to study. One cost-efficient technique for expanding the search to find these very numerous members of the near-Earth object population is to employ very wide field telescopes that cover the entire accessible sky several times each night. This sort of neighborhood watch might be part of the next generation of survey instruments for near-Earth objects.

If we are ever going to deflect an Earth-threatening near-Earth object, we will need to find it well in advance of the predicted strike because any attempt at deflection will likely take several years of preparations. Continuing and expanding our efforts to find and track these objects is critically important for mitigating objects that have our name on them. We need to find them well before they find us. The three most important goals for planetary defense from near-Earth objects is to find them early, find them early, and . . . find them early.

✧ CHAPTER 10 ✧

Deflecting an Earth-Threatening Near-Earth Object

Since hazards from asteroids and comets must apply to inhabited planets all over the Galaxy, if there are such, intelligent beings everywhere will have to unify their home worlds politically, leave their planets, and move small nearby worlds around. Their eventual choice, as ours, is spaceflight or extinction.
—Carl Sagan, *The Pale Blue Dot*

Let's assume that the unimaginable has occurred and a sizable near-Earth asteroid has been discovered on an Earth-threatening trajectory. If there is sufficient time before the predicted impact, there are numerous deflection options that can be considered.

Science nerds love to dream up new asteroid deflection techniques. These include mounting a rocket engine on the surface of an asteroid to push it off its Earth-impact course or affixing a so-called mass driver device to the asteroid's surface to electrostatically throw rocky material off the asteroid in one direction that would provide a small thrust in the opposite direction. Mitigation methods that rely upon surface-mounted devices offer enormous engineering challenges, such as anchoring the device to the asteroid's rough, near-zero gravity surface and then these devices must provide only short bursts of thrusting in one direction as the asteroid rotates to the right position in space. A solar concentrating mirror or a nearby laser device has been suggested because it could burn off, or ablate, asteroid surface material that flies off in one direction, thereby introducing a small thrust in the opposite direction. These methods would also offer significant engineering challenges and the ablating material would likely coat the mirror or laser optics, thus reducing their efficiency. Then there is the fashion designer's favorite mitigation technique of painting the asteroid a different color so the existing Yarkovsky forces pushing on the asteroid as a result of re-radiated solar energy would be modified, thus slowly changing its trajectory to non-impacting. Perhaps a fashionable charcoal gray or an off-white would be suitable.

KISS: Keep It Simple Stupid

While all these options are interesting and fun to think about, we need to focus our attention on the ones that are both simple and technologically viable. These options include precise (but slow) techniques for gradually pulling or pushing the asteroid out of harm's way and somewhat less precise (but more robust) impulsive spacecraft impacts or nuclear detonations. Either of these latter options could instantaneously push upon the asteroid to alter its impact trajectory or completely disrupt the rogue asteroid to disperse all its fragments away from Earth—or at least drastically reduce the individual size and mass of fragments that do collide with Earth.

Deflecting an Asteroid by Running into It

One of the simplest, most mature technologies for deflecting an Earth-threatening asteroid would be to simply run into it with a massive spacecraft. If the asteroid and Earth were predicted to arrive at the same point in space at some future time, the goal would be to alter the asteroid's arrival time by slightly changing its speed. The technology for running into a small celestial body was successfully demonstrated in a spectacular fashion on July 4, 2005, when the Deep Impact spacecraft released a daughter spacecraft that was commanded to run into comet

Figure 10.1. Deep Impact on the nucleus of comet 9P/Tempel 1. After being released from the mother Deep Impact spacecraft, the impactor spacecraft portion of the Deep Impact mission was successfully targeted for the nucleus of comet 9P/Tempel 1 on July 4, 2005. The impact of this 370-kilogram spacecraft on the comet's surface at a relative velocity of 10.3 kilometers per second released thousands of tons of tiny ice and dust particles, which then reflected sunlight that was imaged by the nearby mother spacecraft. After releasing the impacting spacecraft twenty-four hours prior to impact, the mother spacecraft underwent a course correction that allowed it to fly safely past the comet's nucleus at a close approach distance of 500 kilometers. Using three Earth flybys for gravity assists, this spacecraft was renamed EPOXI and went on to fly to within 700 kilometers of short-period comet Hartley 2 on November 4, 2010.
Source: Courtesy of NASA and the University of Maryland.

Tempel 1 while the mother spacecraft sailed on by taking pictures of the impact. Although the nucleus of comet Tempel 1 was six kilometers in extent and thus too massive to show any obvious changes in its trajectory as a result of the impact, the spacecraft navigational skills required for hitting a celestial body with a spacecraft tens of millions of miles away from Earth were demonstrated. The relative velocity of the comet and spacecraft at the time of the collision was ten kilometers per second, about ten times faster than the velocity of a bullet from a high-speed rifle. At this speed, you could travel from Los Angeles to New York City in under seven minutes.

An impulsive shove by a colliding spacecraft could be effective for modest-sized near-Earth asteroids up to a few hundred meters in diameter. For example, a rocky asteroid with a diameter of two hundred meters (about two football fields end to end) that is hit by a five-ton spacecraft at ten kilometers per second could change the asteroid's velocity by more than one centimeter per second. This velocity change means in ten years' time the asteroid's position in its orbit would then be altered by more than 2 Earth radii. So if the asteroid were headed for an Earth collision in ten years, a good thump from an impacting, massive spacecraft would prevent the collision with room to spare. This assumes that the asteroid's orbit was very well-known since it would be prudent to move not only the most likely, or nominal, position of the asteroid off the Earth in ten years' time but all of the possible positions of the asteroid at the time of the predicted collision. That is, in order to sleep easily, we need to move not only the asteroid's nominal position off the Earth in ten years' time but its entire position uncertainty region as well.

If time permits, hitting an Earth-threatening asteroid with a massive spacecraft is a simple and effective means of altering the asteroid's trajectory. However, the technique is somewhat imprecise since the physical characteristics of the asteroid are not likely to be well-known and the push from the impacting spacecraft depends upon the composition, and especially the porosity, of the asteroid itself. If our intrepid spacecraft runs into an asteroid that is rocky with little porosity, the asteroid ejecta from the impact crater will fly back in the direction opposite to the spacecraft's approach direction and these ejecta will increase the push, or momentum, delivered to the asteroid. On the other hand, a porous asteroid would efficiently dissipate the shock, there would be relatively little in the way of ejected material, and the spacecraft hit would cause correspondingly less of a push. If you, with your super-human powers, hurl a rock at an incredible speed toward a cinderblock wall, you're likely to cause far more damage than heaving the same rock at a porous snowbank. I'm not suggesting that either idea is particularly safe but you get the idea. So without first knowing the composition and porosity of the asteroid, it will be tough to determine, ahead of time, just how much the impacting spacecraft will nudge the asteroid. If the first impact doesn't do the trick, hit it again, and yet again if you have the time. But in order to verify that you actually have moved the asteroid the requisite amount, it would be prudent to have another spacecraft already in orbit about, or at least nearby, the asteroid so that the orbiting spacecraft's position, and hence the asteroid's position, could be accurately determined before and after the impacting spacecraft collides with the asteroid. Radio signals sent to and from both spacecraft would be used to accurately determine their positions and trajectories in space. Rusty Schweickart, an Apollo 9 astronaut who has been active in defining near-Earth object issues, has suggested that the spacecraft in residence near the threatening asteroid could verify that an impacting spacecraft actually moved the asteroid the requisite amount and, if necessary and properly designed, could provide the trim maneuvers necessary if the impacting spacecraft didn't quite do the job or if the impulsive push provided by the impactor inadvertently pushed the asteroid into a dynamic keyhole that would allow an Earth impact at a future return to the Earth's neighborhood. One technique for providing this type of trim maneuver would employ the so-called gravity tractor.

The Slow-Pull Gravity Tractor

This novel approach for slowly deflecting an Earth-threatening asteroid would use the gravitational attraction between an asteroid and a nearby thrusting spacecraft to slowly change the asteroid's trajectory. This concept, introduced by astronauts Ed Lu and Stan Love in 2005, envisages a relatively small rotating asteroid and a very close spacecraft that has canted thrusters to avoid the thrust impinging upon the asteroid. Using the gravitational attraction between them as a virtual towline, the spacecraft then acts to either slow down or speed up the asteroid's motion in its orbit. The beauty of this concept is that it doesn't depend upon the porosity, composition, or rotation of the underlying asteroid. Given enough time, the gravity tractor could change the asteroid's trajectory enough so that a predicted Earth impact could be avoided. However, since the mass of even a modest-sized threatening asteroid would dwarf the mass of a nearby gravity tractor spacecraft, the accelerations upon the asteroid are tiny and it would take several weeks of thrusting and then years of coasting at the slightly increased velocity to change the asteroid's orbital position by more than a few kilometers. The gravity tractor concept is most useful for supplying small corrective trim maneuvers after a more energetic impulsive shove is given to the asteroid by, say, a spacecraft impact. The gravity tractor could also be used, if necessary, to nudge the asteroid away from any small dynamic keyholes if there is a close-Earth approach, thus preventing the Earth collision that would follow such a keyhole passage. In addition, the gravity tractor spacecraft, hovering near an Earth-threatening asteroid, could be tracked to precisely determine its position and trajectory, as well as that of the nearby asteroid, before and after a more energetic impact by a massive spacecraft. Hence, while the gravity tractor concept may not be particularly useful for a primary asteroid deflection technique, it could be used subsequent to an impulse by an impacting spacecraft to provide any asteroid corrective trim maneuvers necessary to prevent an Earth collision. At the same time, it could provide the necessary verification that an impulsive shove by a colliding spacecraft was successful.[1]

Nuclear Detonation: Deflection or Fragmentation

While delivering nuclear explosive packages to space is a bit disconcerting and would require careful collaboration with the international

---

[1] The gravity tractor spacecraft would likely be driven using ion-drive engines whereby neutral atoms, like xenon, are first ionized with energetic electrons, then electrostatically accelerated at tremendous velocities, and finally neutralized with an electron source to produce a very small but continuous thrust. An interesting alternate concept to the gravity tractor would direct one of the low-thrust ion engines directly at the asteroid, thus giving it a tiny but continuous push. To keep the spacecraft near the asteroid's surface, an ion engine on the opposite side of the spacecraft would need to thrust with an equal acceleration.

community, the technology is mature and pound for pound, nuclear detonations can deliver the most bang for the available spacecraft mass. There are two nuclear detonation techniques that have been studied by Dave Dearborn at the Lawrence Livermore Radiation Laboratory in Livermore, California. In the first technique, the nuclear device would be exploded just off the asteroid's surface to strongly heat and ablate the asteroid's surface material away from the asteroid in one general direction, thus providing a recoil push in the opposite direction. The most effective type of explosion would be a fusion reaction (hydrogen explosive device) that could generate copious quantities of neutrons that would penetrate the asteroid's surface, strongly heat and vaporize the surface material, and provide the recoil push to alter the asteroid's trajectory. Although initiated by a nuclear detonation, this type of stand-off blast would provide a relatively gentle push upon the asteroid. The total push upon the asteroid would be less than the so-called escape velocity required for surface chunks to escape the asteroid's surface. Hence, very little fragmentation would be expected.[2]

The second nuclear technique assumes the explosive device could be embedded in the asteroid's surface and then detonated in an attempt to completely disrupt the asteroid with the fragments then dispersing through space before reaching the Earth. Dave Dearborn has run computer model simulations to study what it would take to completely fragment an asteroid. According to his computations, a 300-kiloton explosion embedded a few meters into the surface of an asteroid with a diameter of 270 meters (e.g., Apophis) would disrupt the object with most fragments leaving with relative velocities of 20–40 meters per second—well over the escape velocity. The fragments would then disperse along the asteroid's orbital path with only a few percent of the total mass striking the Earth even if the explosion occurred but a few weeks prior to the expected impact. This nuclear option might be considered if the predicted impact were discovered late and there was little

---

[2] As a rule of thumb, the escape velocity from a rocky asteroid, in meters per second, is roughly equal to the asteroid's radius in kilometers. For example, for an asteroid with a radius of 100 meters (0.1 km), its escape velocity would be about 0.1 meters per second, or 10 centimeters per second. That's a very slow 0.2 mph so walking on such an object would be impossible; with a single step, you'd put yourself on an escape trajectory.

time to deal with an expected impactor or if the impactor were too massive to be effectively deflected by one or more spacecraft impacts.

The first nuclear option, or stand-off blast, could be delivered quickly and at high speeds, whereas the embedded surface explosion of the second option would require the extra time for a spacecraft to match trajectories with the asteroid, rendezvous with it, and bury the explosive device in its surface. Implanting a subsurface explosive device at high speeds would not work since current penetrator devices cannot survive impact speeds of more than about one kilometer per second. Thus the stand-off blast would require much less time to carry out than the fragmentation technique.

As with the impulsive push from an impacting spacecraft, the nuclear impulsive options suffer from a poor knowledge of the asteroid's composition and structure so the precise amount of momentum delivered to it would not be well-known. A good deal more work is necessary to better understand the relationship between the deflection impulse and the actual momentum delivered to the asteroid. In other words, for a delivered amount of energy, how will the different compositions and structures of an asteroid react? These types of questions are currently under study using computer simulations by researchers like Keith Holsapple at the University of Washington, Mark Boslough at the Sandia Laboratories in Albuquerque, and Erik Asphaug at the University of California at Santa Cruz. Impact experiments that use expanding gas in guns to propel projectiles at high velocities are used to study the reaction of different types of materials to projectiles of known input energies. These experiments are also necessary to provide empirical data to check the computer simulations. Researchers like Pete Schultz at Brown University and Kevin Housen at the Boeing Company are undertaking these studies.

Then there is the issue of the 1963 Partial Test Ban Treaty and the 1996 Comprehensive Test Ban Treaty that, in principle, prohibit nuclear weapons in outer space. However, these treaties also state that activities in outer space shall be for the common good of all countries and humankind. If a nuclear option were required for mitigating an Earth-threatening body, this principle provides a good legal basis for not limiting mitigation actions that would clearly be for the common good. There is also a fundamental principle of law that if a legal rule leads to "manifest absurdity or unreasonableness," then the rule could be struck down. If a nuclear deflection option were the only way to save the Earth from a significant asteroid strike, treaties banning this would seem to qualify as unreasonable. Even so, the nuclear asteroid deflection option would likely be considered only if all other options were determined inadequate.

MIT Students Saved the World in 1967

Shortly before the MIT spring term in 1967, an announcement of course 16.74, Advanced Space Systems Engineering, appeared on campus bulletin boards. What could be more challenging than a MIT course on advanced space systems engineering? But wait, it gets worse. The course description noted the upcoming close-Earth approach of near-Earth asteroid Icarus on June 14, 1968, and posited that the asteroid would actually hit Earth at that time.[3] The MIT students were asked to establish a plan that must be carried out to stop the impact, the equivalent of 500,000 megatons of TNT explosives—the energy equivalent of one Hiroshima type nuclear blast every second for 445 days! The course description went on to say that if not stopped, this collision would loft 100 million tons of soil and rock into the stratosphere, reduce the sunlight received on Earth's surface, and possibly usher in an ice age. That certainly got the MIT nerd's attention!

The students were told they had but seventy weeks from project inception to Earth impact so there was no time to send a rendezvous spacecraft to the asteroid, which they assumed had a radius of 2,100 feet, or 640 meters. The students broke up into teams to study mission

---

[3] Apollo asteroid (1566) Icarus was discovered on June 26, 1949, by Walter Baade with the Palomar Mountain 48-inch Schmidt telescope. Near the time of its discovery, Icarus passed to within about 15 million kilometers of Earth. On June 14, 1968, it passed within 6.4 million kilometers of Earth and on June 16, 2015, it will pass Earth at an approximate distance of 8 million kilometers. That will remain its closest approach until June 14, 2090, when it will get within about 6.5 million kilometers of Earth. The MIT Icarus project was written up in book form and inspired the (quite forgettable) movie *Meteor*.

plans and deflection options. They determined that Icarus was too massive and too close to the impact time to use anything but a series of six spacecraft, each carrying 100-megaton nuclear devices. Each device would be detonated just off the surface of the asteroid at various times beginning at 72.9 days before impact and winding up with the last detonation at 4.9 days prior to impact. They assumed the explosions would either fragment Icarus or deflect it from its collision course. Each separate launch, using a giant Saturn V launch vehicle, was followed by an intercept monitoring satellite that observed all but the last explosive device from a distance of 1,000 miles behind the impacting spacecraft. I hope these students all got As for their MIT coursework because even though they were studying a topic that has only recently been reexamined, most of what they concluded would still be valid today, more than forty years later.

The Risk Corridor and the Deflection Dilemma

When a potential asteroid impact threat has been identified, there is most often little certainty in the prediction of a possible impact location. The position of the asteroid in space is not perfectly known. The most likely position of the asteroid is centered within a locus of the possible positions. This region is then termed an uncertainty ellipsoid in space, and more often than not it takes the form of a very elongated and very narrow ellipsoid more akin to an enormously long line than a basketball or football. If, in its motion about the Sun, this uncertainty ellipsoid should intersect the Earth, there will be a non-zero impact probability since at least some possible locations of the asteroid could run into Earth. When this long uncertainty ellipsoid, or line, intersects the Earth, it creates a so-called risk corridor where possible impacts could occur. Often the uncertainty ellipsoid is far larger than the Earth itself, but impacts can occur only at the portion that intersects the Earth. As a result of the elongated nature of the uncertainty ellipsoid, the risk corridor across the Earth is most often a very narrow, quite long line. If the asteroid were to strike Earth it would do so only along this risk corridor. Imagine that an Earth-impacting asteroid is predicted

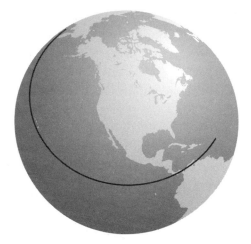

FIGURE 10.2. An example of the so-called risk corridor for a near-Earth asteroid. For an asteroid that has a remote chance of striking Earth, the highly elongated uncertainty ellipsoid region of its orbital position, when projected upon the Earth's surface, forms a narrow "risk corridor" that can wrap nearly around the entire Earth. If the asteroid should actually hit the Earth, it would do so somewhere along this risk corridor.

to hit in the mid-Pacific. If it arrives a bit early, the Earth will not be as far along in its orbit about the Sun so the nominal impact point will move eastward, a bit closer to the leading edge of the Earth. If it arrives a bit late, the nominal impact point will shift westward toward the trailing edge. Deflection attempts to move the asteroid away from an Earth impact would slowly drag the nominal position of the asteroid strike toward either the leading or trailing edge of the Earth. But success could be claimed only if the nominal position of the impact and the totality of the possible impact points were dragged along the risk corridor until completely off the Earth. The impact probability for everyone would then be zero and a global sigh of relief would be heard.

But wait. What if the deflection campaign begins and is then prematurely aborted due to a technical failure before the impact point was dragged completely off the Earth? Now the most likely impact point has been moved from one region of the Earth to another (from the United States to China perhaps?). This has been referred to as the deflection dilemma and it points out rather clearly that asteroid impacts raise serious sociopolitical issues that are every bit as complicated as the technical issues required for deflection. Once a potential impact by a sizable asteroid is predicted, how high must the impact probability rise before action is taken?[4] Who decides what this threshold is, what is the appropriate deflection response, and who carries out the deflection attempt? Who pays if the deflection attempt fails? Clearly international agreements and deflection criteria have to be worked out well in advance of any viable threat.

Who's in Charge?

An asteroid threat is an international problem and hence requires an international solution. There are ongoing discussions of the sociopolitical issues within the Scientific and Technical Committee of the United Nations Committee on the Peaceful Uses of Outer Space (UN COPUOS). Although far from complete, the goal would be to have an internationally approved plan of action already in place before a serious asteroid threat is identified. Presumably some combination of the space-faring nations would be authorized to carry out the deflection campaign.

Most Likely Impact Scenarios

As you consider smaller and smaller near-Earth objects, there are more and more of them. Because there are vastly more small near-Earth objects than there are large ones and since one of these stony objects

---

[4] One of the thorniest sociopolitical issues is when to mount a deflection campaign. Decision makers may find themselves faced with a worrisome but uncertain probability of impact. For example, if a particular near-Earth object, which actually will impact Earth, currently has a 5 percent probability of impact in twenty years' time, does one wait until ground-based observations raise the probability to 50 percent or more before mounting a deflection campaign? If so, the deflection campaign will be far more costly, perhaps impossible, because a larger deflection impulse must be delivered and there is not as much time to deliver the deflection spacecraft to its target. Does one play the odds since the most likely scenario for any non-zero impact probability will have that impact probability drop to zero as additional ground-based observations improve the asteroid's orbit? The impact probability threshold for action is just one issue that needs to be resolved by policymakers within the international community.

would have to be about thirty meters in diameter or larger to cause ground damage, the most likely damaging impact event would be caused by an object of about thirty meters in diameter. There is likely to be more than a million near-Earth asteroids of this size in the Earth's neighborhood, and one would be expected to strike the Earth every few hundred years on average. However, since the trajectories of these objects are uncertain, there would be viable, but ultimately incorrect, asteroid impact threat warnings more often than this. Currently, objects this small are very difficult to discover—less than 1 percent of this population has been discovered to date. Large-aperture, wide-field search telescopes that can scan the entire accessible sky a few times each night would increase the odds that these objects would be discovered at least a few weeks in advance of an impact, but at the moment, the most likely scenario for a damaging asteroid strike will come with little or no warning. Thus civil defense is an important component of asteroid threat mitigation and evacuation away from the risk corridor and perhaps sheltering would seem to be the most common response during anyone's lifetime. However, these small, relatively frequent impactors would do only regional damage whereas the far less frequent but larger impactors could cause global problems. Hence even though the large impactor events are extremely rare, it is this group that is by far the greatest threat for causing loss of life when averaged over very long time periods.

We Found an Asteroid Impactor—Whom Do We Call?

The necessary technology for finding and tracking near-Earth asteroids is fairly well developed and there are at least plausible plans for the technology necessary for deflecting an asteroid if there is sufficient time to do so. Within the NASA organization, there are established guidelines for whom should be notified when a credible threat is identified. However, internationally, plans are only in the formative stage in terms of answering thorny questions such as these: Who would be given the authority to act on behalf of all humankind? How would such an effort be coordinated, funded and carried out? Could civil defense plans be established for what might include evacuations of a large region that may include several countries?

Summary

This book has made an effort to stress the importance of near-Earth asteroids and comets in terms of science, the origin and development of life, future space resources, and the defense of our planet from a horrific natural disaster. Near-Earth objects are the leftover bits and pieces from the early solar system formation process and they are among the least-changed members of that system. Hence these comets and asteroids offer clues to the chemical composition and thermal environment under which our solar system formed 4.6 billion years ago. Following the creation of the early Earth, a rain of these objects likely delivered a veneer of carbon-based materials and water to the Earth's surface, thus providing the building blocks of life. Once life did form, infrequent collisions by large asteroids or comets created extinction events and punctuated evolution, allowing only the most adaptable species to evolve further. We humans may owe our very existence and our current position atop the food chain to these objects. Rich in minerals, metals, and water resources, near-Earth objects could one day provide the raw materials for building interplanetary habitats. Their water resources could be used to support life and provide rocket fuel when the water is broken down into hydrogen and oxygen. Near-Earth objects may one day be the fueling stations and watering holes for interplanetary exploration. Ironically, the easiest ones to reach and mine are also those that are most likely to one day collide with Earth and perhaps disrupt or destroy our fragile civilization. We need to find them early and track them to ensure that none among them has our name on it. While these objects are critically important for our future, if we don't find them before they find us, we may not even have a future.

Near-Earth objects are among the smallest members of the solar system, but their diminutive size is in no way proportional to their importance. When it comes to their role in the development and future of humankind, next to the Sun itself, theirs is the most important realm.

✦ REFERENCES ✦

Preface

Lovell, E. J. Jr., ed. Medwin's *"Conversations of Lord Byron."* Princeton: Princeton University Press, 1966.

Chapter 2

Campins, H., K. Hargrove, N. Pinilla-Alonso, E. Howell, M. S. Kelley, J. Licandro, T. Mothé-Diniz, Y. Fernández, and J. Ziffer. "Water Ice and Organics on the Surface of the Asteroid 24 Themis." *Nature* 464 (2010): 1320–21.

Fernández, J. A. "On the Existence of a Comet Belt beyond Jupiter." *Monthly Notices of the Royal Astronomical Society* 192 (1980): 481–91.

Jewitt, D. "What Else Is out There?" *Sky and Telescope* 119 (2010): 20–24.

Rivkin, Andrew S., and Joshua P. Emery. "Detection of Ice and Organics on an Asteroid surface." *Nature* 464 (2010): 1322–23.

Stern, Alan, "Secrets of the Kuiper belt." *Astronomy* (April 2010): 30–35.

Chapter 3

Chesley, S. R., S. J. Ostro, D. Vokrouhlický, D. Capek, J. D. Giorgini, M. C. Nolan, J.-L. Margot, A. A. Hine, L.A.M. Benner, and A. B. Chamberlin. "Direct Detection of the Yarkovsky Effect by Radar Ranging to Asteroid 6489 Golevka." *Science* 302 (2003): 1739–42.

Fernández, J. A., and W.-H Ip. "Some Dynamical Aspects of the Accretion of Uranus and Neptune: The Exchange of Orbital Angular Momentum with Planetesimals." *Icarus* 58, no. 1 (1984): 109–20.

Gomes, R., H. F. Levison, K. Tsiganis, and A. Morbidelli. "Origin of the Cataclysmic Late Heavy Bombardment Period of the Terrestrial Planets." *Nature* 435 (2005): 466–69.

Levison, H. F., M. J. Duncan, R. Brasser, and D. E. Kaufmann. "Capture of the Sun's Oort Cloud from Its Birth Cluster." *Science* 329 (2010): 187–90.

Malhotra, R. "The Origin of Pluto's Orbit: Implications for the Solar System beyond Neptune." *Astronomical Journal* 100, no. 1 (1995): 420–29.

Morbidelli, A., H. F. Levison, K. Tsiganis, and R. Gomes. "Chaotic Capture of Jupiter's Trojan Asteroids in the Early Solar System." *Nature* 435 (2005): 462–65.

Tsiganis, K. R., R. Gomes, A. Morbidelli, and H. F. Levison. "Origin of the Orbital Architecture of the Giant Planets of the Solar System." *Nature* 435 (2005): 459–61.

Chapter 4

Alvarez, L. W., W. Alvarez, F. Asaro, and H. V. Michel. "Extraterrestrial Cause for the Cretaceous-Tertiary Extinction." *Science* 208 (1980): 1095–1108.
Hildebrand, A. R., G. T. Penfield, D. A. Kring, M. Pilkington, A. Camargo Z., S. B. Jacobsen, and W. V. Boynton. "Chicxulub Crater: A Possible Cretaceous/ Tertiary Boundary Impact Crater on the Yucatán Peninsula, Mexico." *Geology* 19 (1991): 867–71.

Chapter 6

Chapman, Clark R., and David Morrison. "Impacts on the Earth by Asteroids and Comets: Assessing the Hazard." *Nature* 367 (1994): 33–40.
Bottke, W. F., Jr., A. Cellino, P. Paolicchi, and R. P. Binzel. "An Overview of the Asteroids: The Asteroids III Perspective." In *Asteroids III*, ed. W. F. Bottke Jr., A. Cellino, P. Paolicchi, and R. P. Binzel, 3–15. Tucson: University of Arizona Press, 2002.
Davis, D. R., C. R. Chapman, R. Greenberg, S. J. Weidenschilling, and A. W. Harris. "Collisional Evolution of Asteroids: Populations, Rotations and Velocities." In *Asteroids*, ed. T. Gehrels, 528–57. Tucson: University of Arizona Press, 1979.
Ostro, S. J., R. S. Hudson, M. C. Nolan, J.-L. Margot, D. J. Scheeres, D. B. Cambell, C. Magri, J. D. Giorgini, and D. K. Yeomans. "Radar Observations of Asteroid 216 Kleopatra." *Science* 288 (2000): 836–39.

Chapter 7

Landis, Rob. "NEOs Ho! The Asteroid Option." *Griffith Observer* 73, no. 5 (2009): 3–19.
Lewis, John S., *Mining the Sky*. Reading, MA: Helix Books, Addison-Wesley, 1997.

Chapter 8

Boslough, M., and D. Crawford. "Low-Altitude Airbursts and the Impact Threat." *International Journal of Impact Engineering* 35 (2008): 1441–48.
Halliday, Ian, A. T. Blackwell, and A. A. Griffin. "Meteorite Impacts on Humans and Buildings." *Nature* 318 (November 28, 1985): 317.
Harris, Alan. "What Spaceguard Did." *Nature* 453 (June 26, 2008): 1178–79.
Lloyd, Robin. "Competing Catastrophes: What's the Bigger Menace, an Asteroid Impact or Climate Change?" *Scientific American*, March 31, 2010.
National Research Council. *Defending Planet Earth: Near-Earth Object Surveys and Hazard Mitigation Strategies*. Washington, DC: National Academies Press, 2010.

Sekanina, Z., and D. K. Yeomans. "Close Encounters and Collisions of Comets with the Earth." *Astronomical Journal* 89 (1984): 154–61.

Study to Determine the Feasibility of Extending the Search for Near-Earth Objects to Smaller Limiting Diameters: Report of the Near-Earth Object Science Definition Team. August 22, 2003.

Toon, O. B., K. Zahnle, D. Morrison, R. P. Turco, and C. Covey. "Environmental Perturbations Caused by the Impact of Asteroids and Comets." *Reviews of Geophysics* 35 (1997): 41–78.

Van Dorn, W. G., B. LeMehaute, and L.-S. Hwant. *Handbook of Explosion-Generated Water Waves.* Vol. 1, *State of the Art.* Pasadena, CA: Tetra Tech, 1968.

Chapter 9

Giorgini, J. D., S. J. Ostro, L.A.M. Benner, P. W. Chodas, S. R. Chesley, R. S. Hudson, M. C. Nolan, A. R. Klemola, E. M. Standish, R. F. Jurgens, R. Rose, A. B. Chamberlin, D. K. Yeomans, and J.-L. Margot. "Asteroid 1950 DA's Encounter with Earth in 2880: Physical Limits of Collision Probability Prediction." *Science* 296 (April 5, 2002): 132–36.

Jenniskens, P., M. H. Shaddad, D. Numan, S. Elsir, A. M. Kudoda, M. E. Zolensky, L. Le, G. A. Robinson, J. M. Friedrich, D. Rumble, A. Steele, S. R. Chesley, A. Fitzsimmons, S. Duddy, H. H. Hsieh, G. Ramsay, P. G. Brown, W. N. Edwards, E. Tagliaferri, M. B. Boslough, R. E. Spalding, R. Dantowitz, M. Kozubal, P. Pravec, J. Borovicka, Z. Charvat, J. Vaubaillon, J. Kuiper, J. Albers, J. L. Bishop, R. L. Mancinelli, S. A. Sanford, S. N. Milam, M. Nuevo, and S. P. Worden. "The Impact and Recovery of Asteroid 2008 TC3." *Nature* 458 (March 26, 2009): 485–88.

Milani, A., S. R. Chesley, M. E. Sansaturio, F. Bernardi, G. B. Valsecchi, and O. Arratia. "Long-Term Impact Risk for (101955) 1999 RQ36." *Icarus* 203, no. 2 (2009): 460–71.

Chapter 10

Lu, E. T., and S. G. Love. "A Gravitational Tractor for Towing Asteroids." *Nature* 438, no. 2 (2005): 177–78.

National Research Council. *Defending Planet Earth: Near-Earth Object Surveys and Hazard Mitigation Strategies.* Washington, DC: National Academies Press, 2010.

Project Icarus. *MIT Student Project in Systems Engineering.* Cambridge, MA: MIT Press, 1968.

Schweickart, R. L., T. D. Jones, F. von der Dunk, S. Camacho-Lara, and Association of Space Explorers International Panel on Asteroid Threat Mitigation. *Asteroid Threats: A Call for Global Response.* Houston, TX: Association of Space Explorers, 2008.

✦ INDEX OF ASTEROID AND COMETARY OBJECTS ✦